MATERIAL SCIENCE & ENGINEERING

AUTHORS

Dr. Vinod Kumar
Assistant Professor and Former Head,
Department of Metallurgical Engineering and Materials Science, IIT Indore,
Simrol, Khandwa Road, Indore -453552,
Madhya Pradesh.

Dr. Nisheeth Kumar Prasad
Assistant Professor,
Department of Metallurgical Engineering and Materials Science, IIT Indore,Simrol,
Khandwa Road, Indore-453552.

REVIEWER

Dr. Sirisha Uppaluri
Associate Professor,
Department of Civil Engineering
Christ (Deemed To Be University), Bengaluru

KHANNA BOOK PUBLISHING CO. (P) Ltd.
4C/4344, Ansari Road, Darya Ganj,
New Delhi, India

All India Council for Technical Education
Nelson Mandela Marg, Vasant Kunj,
New Delhi, India

Dear Readers,

To prevent the piracy, this book is secured with HIGH SECURITY HOLOGRAM on the front title cover. In case you don't find the hologram on the front cover title, please write us to at contact@khannabooks.com or whatsapp us at +91-99109 09320 and avail special gift voucher for yourself.

Specimen of Hologram on front Cover title:

Moreover, there is a SPECIAL DISCOUNT COUPON for you with EVERY HOLOGRAM.

How to avail this SPECIAL DISCOUNT:

Step 1: Scratch the hologram

Step 2: Under the scratch area, your "coupon code" is available

Step 3: Logon to www.khannabooks.com

Step 4: Use your "coupon code" in the shopping cart and get your copy at a special discount

Step 5: Enjoy your reading!

ISBN: 978-93-55383-87-7

Material Science & Engineering

by Dr. Vinod Kumar & Dr. Nisheeth Kumar Prasad

First Edition: 2025

Originally Published by:
All India Council for Technical Education, Ministry of Education, GOI

Exclusively Published & Distributed by:
Khanna Book Publishing Co. (P) Ltd.
4C/4344, Ansari Road, Daryaganj,
New Delhi (INDIA)

Visit us at: **www.khannabooks.com**
Write us at: **contact@khannabooks.com**

To view complete list of books, Please scan the QR Code:

KPH

Printed in Bharat.

प्रो. टी. जी. सीताराम
अध्यक्ष
Prof. T. G. Sitharam
Chairman

सत्यमेव जयते

आज़ादी का अमृत महोत्सव

अखिल भारतीय तकनीकी शिक्षा परिषद्
(भारत सरकार का एक सांविधिक निकाय)
(शिक्षा मंत्रालय, भारत सरकार)
नेल्सन मंडेला मार्ग, वसंत कुंज, नई दिल्ली-110070
दूरभाष : 011-26131498
ई-मेल : chairman@aicte-india.org

ALL INDIA COUNCIL FOR TECHNICAL EDUCATION
(A STATUTORY BODY OF THE GOVT. OF INDIA)
(Ministry of Education, Govt. of India)
Nelson Mandela Marg, Vasant Kunj, New Delhi-110070
Phone : 011-26131498
E-mail : chairman@aicte-india.org

FOREWORD

Engineers are the backbone of any modern society. They are the ones responsible for the marvels as well as the improved quality of life across the world. Engineers have driven humanity towards greater heights in a more evolved and unprecedented manner.

The All India Council for Technical Education (AICTE), have spared no efforts towards the strengthening of the technical education in the country. AICTE is always committed towards promoting quality Technical Education to make India a modern developed nation emphasizing on the overall welfare of mankind.

An array of initiatives has been taken by AICTE in last decade which have been accelerated now by the National Education Policy (NEP) 2020. The implementation of NEP under the visionary leadership of Hon'ble Prime Minister of India envisages the provision for education in regional languages to all, thereby ensuring that every graduate becomes competent enough and is in a position to contribute towards the national growth and development through innovation & entrepreneurship.

One of the spheres where AICTE had been relentlessly working since past couple of years is providing high quality original technical contents at Under Graduate & Diploma level prepared and translated by eminent educators in various Indian languages to its aspirants. For students pursuing 2nd year of their Engineering education, AICTE has identified 87 books, which shall be translated into 12 Indian languages - Hindi, Tamil, Gujarati, Odia, Bengali, Kannada, Urdu, Punjabi, Telugu, Marathi, Assamese & Malayalam. In addition to the English medium, books in different Indian Languages are going to support the students to understand the concepts in their respective mother tongue.

On behalf of AICTE, I express sincere gratitude to all distinguished authors, reviewers and translators from the renowned institutions of high repute for their admirable contribution in a record span of time.

AICTE is confident that these outcomes based original contents shall help aspirants to master the subject with comprehension and greater ease.

(Prof. T. G. Sitharam)

ACKNOWLEDGEMENT

The authors are grateful to the authorities of AICTE, particularly Prof. T. G. Sitharam, Chairman; Dr. Abhay Jere, Vice-Chairman; Prof. Rajive Kumar, Member-Secretary; and Dr. Ramesh Unnikrishnan, Advisor-II and Dr. Sunil Luthra, Director, Training and Learning Bureau for their planning to publish the book on Material Science & Engineering. We sincerely acknowledge the valuable contributions of the reviewer of the book Prof. Parvati Ramaswamy, Ex-Professor, Department of Mechanical and Automobile Engineering, Christ University, Bangalore for making it students' friendly. Further, we would like to thank our Ph.D. student Mr. Gokul Pillai for helping us in making schematics in few chapters.

This book is an outcome of various suggestions of AICTE members, experts and authors who shared their opinion and thought to further develop the engineering education in our country. Acknowledgements are due to the contributors and different workers in this field whose published books, review articles, papers, photographs, footnotes, references, and other valuable information enriched us at the time of writing the book.

The author, Vinod Kumar, would like to acknowledge the encouragement, support and everlasting love of his father, Gangadhar Thakur, throughout his life and professional career. Authors dedicate this book to their parents.

Dr. Vinod Kumar

Dr. Nisheeth Kumar Prasad

PREFACE

The book titled "Material Science & Engineering" is an outcome of the rich experience of our teaching of basic metallurgical and materials engineering courses. The initiation of writing this book is to expose the students of engineering diploma to the fundamental topics of materials engineering as well as enable them to get an insight of the subject. Keeping in mind the purpose of wide coverage as well as to provide essential supplementary information, we have included the topics recommended by AICTE, in a very systematic and orderly manner throughout the book. Efforts have been made to explain the fundamental concepts of the subject in the simplest possible way.

During the process of preparation of the manuscript, we have considered the various standard textbooks and accordingly we have developed sections like critical questions, solved and supplementary problems etc. While preparing the different sections emphasis has also been laid on definitions, pictorial representation and on comprehensive synopsis of important questions for a quick revision of the basic principles. The book covers all types of medium and advanced level problems, and these have been presented in a very logical and systematic manner. The gradations of those problems have been tested over many years of teaching to a wide variety of students.

Apart from illustrations and examples as required, we have enriched the book with numerous solved problems in every unit for proper understanding of the related topics. It is important to note that at the end of each chapter, we have included the relevant laboratory practical that will further enhance the learning of the student on the subject. In addition, some essential information is also provided for the users under the heading "Know More" at the end of each chapter.

As far as the present book is concerned, "Material Science & Engineering" is meant to provide a thorough understanding of the important concepts of materials engineering to the diploma and engineering students, which includes a broad spectrum of topics related to this field. This book will prepare the students to apply the acquired knowledge in the practical field and be able to tackle such challenges in industries. The subject matters are presented in a constructive manner to prepare engineering diploma students to work in different sectors or in national laboratories at the very forefront of technology.

We sincerely hope that the book will inspire the students to learn and discuss the ideas behind basic principles of Material Science and Engineering and will surely contribute to the development of a solid foundation of the subject. We would be thankful to all beneficial comments and suggestions which will contribute to the improvement of the future editions of the book. It gives us immense pleasure to place this book in the hands of the teachers and students. It was indeed a big pleasure to work on different aspects covered in the book.

Dr. Rinku Kumar

Dr. Sandeep Panchal

OUTCOME BASED EDUCATION

For the implementation of an outcome-based education the first requirement is to develop an outcome-based curriculum and incorporate an outcome-based assessment in the education system. By going through outcome-based assessments, evaluators will be able to evaluate whether the students have achieved the outlined standard, specific and measurable outcomes. With the proper incorporation of outcome-based education there will be a definite commitment to achieve a minimum standard for all learners without giving up at any level. At the end of the programme running with the aid of outcome-based education, a student will be able to arrive at the following outcomes:

Programme Outcomes (POs) are statements that describe what students are expected to know and be able to do upon graduating from the program. These relate to the skills, knowledge, analytical ability attitude and behaviour that students acquire through the program. The POs essentially indicate what the students can do from subject-wise knowledge acquired by them during the program. As such, POs define the professional profile of an engineering diploma graduate.

National Board of Accreditation (NBA) has defined the following seven POs for an Engineering diploma graduate:

PO-1. Basic and Discipline specific knowledge: Apply knowledge of basic mathematics, science and engineering fundamentals and engineering specialization to solve the engineering problems.

PO-2. Problem analysis: Identify and analyses well-defined engineering problems using codified standard methods.

PO-3. Design/ development of solutions: Design solutions for well-defined technical problems and assist with the design of systems components or processes to meet specified needs.

PO-4. Engineering Tools, Experimentation and Testing: Apply modern engineering tools and appropriate techniques to conduct standard tests and measurements.

PO-5 Engineering practices for society, sustainability and environment: Apply appropriate technology in context of society, sustainability, environment and ethical practices.

PO-6. Project Management: Use engineering management principles individually, as a team member or a leader to manage projects and effectively communicate about well-defined engineering activities.

PO-7. Life-long learning: Ability to analyse individual needs and engage in updating in the context of technological changes.

COURSE OUTCOMES

By the end of the course the students are expected to learn:

CO-1: The importance of engineering materials in day-to-day life and underlying principles which describe the properties of such materials.

CO-2: The role and application of phase diagram in materials selection and design. Also, the importance of ferrous alloys in structural applications.

CO-3: About various commercial non-ferrous alloys, designation, phase diagrams and their applications.

CO-4: The failure of engineering materials and various destructive and non-destructive test methods along with their standards.

CO-5: The significance of corrosion in materials degradation and different methods to control corrosion.

Mapping of Course Outcomes with Programme Outcomes to be done according to the matrix given below:

Course Outcomes	EXPECTED MAPPING WITH PROGRAMME OUTCOMES (1- Weak Correlation; 2- Medium correlation; 3- Strong Correlation)						
	PO-1	PO-2	PO-3	PO-4	PO-5	PO-6	PO-7
CO-1	3	2	3	2	3	1	3
CO-2	3	3	3	2	2	2	3
CO-3	3	3	3	2	2	2	2
CO-4	3	3	3	3	3	3	3
CO-5	3	2	3	3	3	2	3

GUIDELINES FOR TEACHERS

To implement Outcome Based Education (OBE) knowledge level and skill set of the students should be enhanced. Teachers should take a major responsibility for the proper implementation of OBE. Some of the responsibilities (not limited to) for the teachers in OBE system may be as follows:

- Within reasonable constraint, they should manoeuvre time to the best advantage of all students.
- They should assess the students only upon certain defined criterion without considering any other potential ineligibility to discriminate them.
- They should try to grow the learning abilities of the students to a certain level before they leave the institute.
- They should try to ensure that all the students are equipped with the quality knowledge as well as competence after they finish their education.
- They should always encourage the students to develop their ultimate performance capabilities.
- They should facilitate and encourage group work and team work to consolidate newer approach.
- They should follow Blooms taxonomy in every part of the assessment.

Bloom's Taxonomy

Level	Teacher should Check	Student should be able to	Possible Mode of Assessment
Creating	Students ability to create	Design or Create	Mini project
Evaluating	Students ability to Justify	Argue or Defend	Assignment
Analysing	Students ability to distinguish	Differentiate or Distinguish	Project/Lab Methodology
Applying	Students ability to use information	Operate or Demonstrate	Technical Presentation/ Domonstration
Understanding	Students ability to explain the ideas	Explain or Classify	Presentation/Seminar
Remembering	Students ability to recall (or remember)	Define or Recall	Quiz

GUIDELINES FOR STUDENTS

Students should take equal responsibility for implementing the OBE. Some of the responsibilities (not limited to) for the students in OBE system are as follows:

- Students should be well aware of each UO before the start of a unit in each and every course.
- Students should be well aware of each CO before the start of the course.
- Students should be well aware of each PO before the start of the programme.
- Students should think critically and reasonably with proper reflection and action.
- Learning of the students should be connected and integrated with practical and real life consequences.
- Students should be well aware of their competency at every level of OBE.

ABBREVIATIONS AND SYMBOLS

List of Abbreviations

General Terms			
Abbreviations	**Full form**	**Abbreviations**	**Full form**
LRO	Long range order	HSS	High speed steel
SRO	Short range order	PH	Precipitation hardening
MMC	Metal matrix composite	SCC	Stress corrosion cracking
PMC	Polymer matrix composite	CDA	Copper development association
CMC	Ceramic matrix composite	SFE	Stacking fault energy
IC	Integrated circuits	UTS	Ultimate tensile strength
BCC	Body centered cubic	BHN	Brinel hardness number
FCC	Face centered cubic	NDT	Non-destructive testing
BCT	Body centered tetragonal	MPI	Magnetic particle inspection
HCP	Hexagonal close packed	NACE	National Association of Corrosion Engineers
CCP	Cubic close packing		
SG	Spheroidal graphite	TS	Thermal spray
CNT	Carbon nanotubes	D-gun	Detonation gun
BF	Blast furnace	HVOF	High velocity oxyfuel
BOF	Basic oxygen furnace	ICP	Induction coupled plasma
EAF	Electric arc furnace	PVD	Physical vapor deposition
BIS	Bureau of Indian Standards	CVD	Chemical vapor deposition
IS	Indian standard		
AISI	American Iron and Steel Institute		
SAE	SocietyforAutomotiveEngineers		

List of Units and Symbols

Symbols	Description	Symbols	Description
a-Po	Polonium	γs	Surface energy per unit area
a	Lattice parameter	γP	energy required for plastic deformation per unit area of the crack
F	angle between two Miller indices	σm	mean stress
d(hkl)	interplanar spacing	σr	range of stress

Symbols	Description	Symbols	Description
N	Avogadro number (6.023 × 1023)	σa	stress amplitude
	enthalpy of formation of single vacancy in a crystal (kJ/mol)	R	stress ratio
(, (), and ()	Allotropic forms of iron	τ	shear stress
	Maximum stress at crack tip	MT	Tortional moment
σ	Applied tensile stress normal to crack length	J	polar moment of inertia
UE	Elastic strain energy		

LIST OF FIGURES

LIST OF TABLES

CONTENTS

1 Crystal Structures and Bonds

UNIT SPECIFICS

Through this unit the following aspects of crystal structures and different types of bonds have been discussed:

- Introduction to material science and engineering: Classification of engineering materials.
- Crystal system: The seven basic crystal systems; Crystal structure for metallic elements: BCC,
- FCC and HCP; Coordination number for Simple Cubic, BCC and FCC; Atomic radius: definition, atomic radius for Simple Cubic, BCC and FCC; Atomic Packing Factor for Simple Cubic, BCC, FCC and HCP; Simple problems on finding number of atoms for a unit cell.
- Bonds in solids: Classification - primary or chemical bond, secondary or molecular bond; Types of primary bonds: Ionic, Covalent and Metallic Bonds; Types of secondary bonds: Dispersion bond, Dipole bond and Hydrogen bond.

The fundamental concepts of crystal structure have been discussed in this text to convey the significance of the topics to the students. The different types of crystal structures and bonding have been explained along with the examples and their importance in case of engineering materials.

Besides giving many multiple-choice questions as well as questions of short and long answer types marked in two categories by following lower and higher order of Bloom's taxonomy, a list of references and suggested readings are given in the unit so that one can go through them for further advanced knowledge.

This section also highlights the examples of some applications of the subject matter for industrial applications on a variety of aspects and/or our day-to-day real life (whichever applicable), and finally inquisitiveness associated with the topics of the unit.

RATIONALE

This unit helps the students to have fundamental understanding about the crystal structures and bonding. After studying this unit, you will be able to

- Understand the importance of materials science and engineering
- Define general characteristics of solid materials

- Classify engineering materials (based on their nature, area of their uses, and levels of structure)
- Define space lattice, motif, crystal and unit cell
- Describe close packed structure and different types of voids
- Calculate the packing factor of different types of cubic unit cells
- Classify the imperfections in solids
- Understand the effect of imperfections in crystalline materials on their properties
- Explain the formation of different types of bonds in solids

PRE-REQUISITES

Physical chemistry and solid-state physics.

UNIT OUTCOMES

List of outcomes of this unit is as follows:

U1-O1: To know the importance of material science and engineering.

U1-O2: To have a basic understanding about general characteristics of solid materials.

U1-O3: To have knowledge about various crystal structures, unit cell and Miller Indices.

U1-O4: To know about the various bonds in solids.

Unit-1 Outcomes	EXPECTED MAPPING WITH COURSE OUTCOMES (1- Weak Correlation; 2- Medium correlation; 3- Strong Correlation)				
	CO-1	CO-2	CO-3	CO-4	CO-5
U1-O1	3	2	2	2	2
U1-O2	3	2	2	2	1
U1-O3	3	2	2	2	1
U1-O4	3	2	2	2	1

1.1 INTRODUCTION TO MATERIAL SCIENCE AND ENGINEERING

Material science is related to the basic understanding of the structure, processing and properties of materials. Material engineering primarily deals with the fundamental and applied knowledge of materials to develop useful products needed for an industry or society. Thus, material science and engineering is mainly dealt with the combined knowledge gained by understanding the structure, processing, properties and performance of engineering materials. Fig. 1.1 shows a diagram to illustrate the wider scope of material science and engineering.

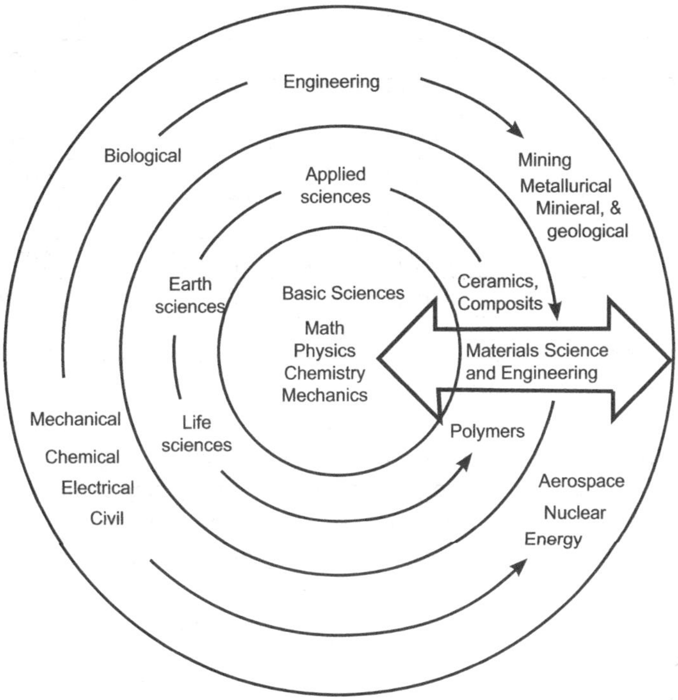

Fig. 1.1: Schematic illustrating scope of material science and engineering

In the current scenario, material science and engineering forms a link between basic science (including mechanics) and engineering disciplines. The scope of material science and engineering is very much relevant to almost every industry.

1.2 GENERAL CHARACTERISTICS OF SOLID MATERIALS

The elements are also materials. Materials in elemental form are relatively scarce in the grand scheme of items. Table 1.1 shows that most of our universe is made up of helium and hydrogen whereas the earth crust is made up of several elements. The earth's crust is dominated by oxygen and silicon.

Table 1.1: Relative Abundance of Elements in universe and in earth's crust

Name of Element (symbol)	Relative abundance w.r.to Si (Si = 1)	Relative abundance in earth's crust (ppm)
Hydrogen (H)	12,000 (1st)	1,400 (10th)
Helium (He)	2,800 (2nd)	0.003
Oxygen (O)	16 (3rd)	**466,000 (1st)**
Nitrogen (N)	8 (4th)	50
Carbon (C)	3	320
Iron (Fe)	2.6	50,000 (4th)
Silicon (Si)	1	**277,200 (2nd)**
Magnesium (Mg)	0.89	20,900 (8th)

Sulphur (S)	0.33	520
Nickel (Ni)	0.21	80
Aluminium (Al)	0.09	**81,300 (3rd)**
Calcium (Ca)	0.07	36,300 (5th)
Sodium (Na)	0.045	28,300 (6th)
Chlorine (Cl)	0.025	314
Potassium (K)	Very less	25,900 (7th)
Titanium (Ti)	Very less	4,400 (9th)

As we know, a material is a matter that has weight and occupies space. Matter can exist in four fundamental states namely, solid, liquid, gas, and plasma. In our surroundings, we observe matters either in the form of solid, liquid or gas. Plasma contains a significant amount of charged particles. Lightening is a common example of partial ionized plasma. Plasma technology is being used in several applications, such as plasma display TV panel, sputtering, mass spectroscopy (using Inductively coupled plasma), semiconductor device fabrication, etc. However, discussing the characteristics and classification of plasma is beyond the scope of this book. Thus, only three states of matter namely solid, liquid, and gas will be discussed in this book. Thermodynamic parameters will decide which of these three states would be the most stable state of any given substance. There are two major factors which describe the stability of the substance at a given temperature and pressure:

(a) Intermolecular forces

(b) Thermal energy

Intermolecular forces tend to retain the atoms (or ions, or molecules) closer, whereas opposite factors i.e. thermal energy tends to keep them apart. The most stable state of a matter depends on the net effect of these two factors.

At very high temperatures, the atoms (or ions, or molecules) have enough thermal energy so that the effect of intermolecular forces is small and as a result the substance acquires a gaseous state. Thus, the gaseous state of a typical matter is more stable at very high temperature.

At moderate temperature, the atoms (or ions, or molecules) have sufficient thermal energy as compared to intermolecular forces to move them relative to each other and the substance acquires liquid state. Thus, the liquid state of a typical matter is more stable at a very moderate temperature.

At sufficiently low temperature, the atoms (or ions, or molecules) have low thermal energy and the intermolecular forces dominate the state. In this condition, the intermolecular forces bring constituent particles sufficiently close and hence they occupy fixed positions. However, they can oscillate about their mean position even at sufficiently low temperatures and the substance exists in solid state. The substance keeps definite mass in case of all states.

The substance in solid state has following characteristic properties:

(i) Strong intermolecular forces as compared to thermal energy.

(ii) Definite shape and volume.

(iii) Short intermolecular distances

(iv) The constituent particles (atoms, ions, or molecules) occupy fixed positions and can only oscillate about their mean positions under given temperature and pressure. Hence, they are commonly rigid and incompressible.

1.2.1 CLASSIFICATION OF SOLIDS

Solid materials can be broadly classified into two groups namely, crystalline and non-crystalline (i.e. amorphous). This classification is based on the nature of order present in the arrangement of constituent particles (atoms, ions, or molecules) of solids.

Crystalline solids: If the constituent particles of a solid are arranged in a specific pattern that repeats itself in three dimensions and has a *long-range order (LRO)* then that solid is commonly called as *crystalline solid.* Thus, a *crystalline solid* has a long-range periodicity characteristic which means the regular pattern of arrangement of constituent particles repeats itself periodically over the entire crystal. Metals, alloys, and few minerals / ceramic materials are typical examples of crystalline solids. Typical examples of crystalline solid materials are summarised in Table 1.2. The key features of crystalline materials are tabulated in Table 1.3.

A crystalline solid may be either a single crystal or aggregates of many crystals. It consists of a large number of well-defined small crystals which have a definite characteristic geometrical shape. Solid having aggregates of various crystals separated by boundaries is referred to as a polycrystalline solid. Metals and their alloys are generally crystalline and are made up of small crystals or grains. Crystalline solids will remain crystalline even when the solid substance is either broken into nano-sized grains or used to manufacture a large component.

Table 1.2: Typical examples of crystalline solid materials

Type	Examples	Remarks
Metals	Fe, Au, Ag, Cu, Mg, Ti, U, Li, Si, Ga, etc.	Every metal has specific characteristics.
Alloys	Steels, brass, bronze, duralumin, etc.	Alloys are used for making useful products needed for an industry or society.
Minerals / Ceramics	Sodium chloride (NaCl)	Halite known as rock salt is a typical example.
	Pyrite (FeS_2)	Due to its appearance (brassy yellow) it is called fool's gold.
	Amethyst (SiO_2)	Quartz (SiO_2) changes from Colourless to black through various colours.

Non-crystalline (Amorphous) solids: In contrast to crystalline materials, amorphous solids have only *short-range order (SRO)* and the constituent particles are not arranged in a periodic and repeatable manner. The key differences are summarised in Table 1.3.

This implies that the ordering in amorphous solids exist only in the immediate neighbourhood of the atoms/ions/molecules. An amorphous solid consists of particles of irregular shape. Glass, plastics, rubber, metallic glass, and amorphous silicon are typical examples of amorphous solids. The number and kind of nearest neighbours that a constituent particle has in a solid state, is nearly the same for both the crystalline and amorphous forms of the solid. However, due to the differences in the arrangement of the constituent particles and their individual characteristics, the two types of solids differ in their properties.

Table 1.3: Key distinction between crystalline and amorphous solids

Property	Crystalline solids	Amorphous Solids
Shape of constituent particles	Definite and regular characteristic geometrical shape	Irregular shape
Order in arrangement of constituent particles	Long range order	Only short-range order
Nature of solid	True solid	Supercooled liquid or pseudo-solid
Melting point	Sharp melting point at a fixed temperature	Soften over a range of temperatures before melting.
Heat of fusion	Possess definite and characteristic heat of fusion	Do not have definite heat of fusion
Anisotropy	Show different properties along different directions in the same crystal	Mostly isotropic in nature
Cleavage characteristics when cut with a sharp edge tool	Split into two pieces having plain and smooth surfaces	Cut into two pieces having irregular surfaces
X-ray diffraction pattern	Diffract X-rays and produce a distinctive X-ray diffraction pattern	Diffract X-rays poorly and do not produce a distinctive X-ray diffraction pattern

1.3 CLASSIFICATION OF ENGINEERING MATERIALS

Engineering materials are those types of materials which are being used for one or more types of engineering applications. We have learnt in earlier classes that materials can be conductors, semiconductors or insulators based on their conductivities. We have also studied that all metals are conductors. Currently, more than 50,000 types of materials are available and hence their classification is essential to understand their characteristics.

1.3.1 Classification of Engineering Materials based on their Nature

Conventionally most engineering materials are divided into three main groups: (a) metals and alloys, (b) polymers, and (c) ceramics based on chemical nature and atomic structure. In this unit we shall distinguish among them based on some of their engineering importance such as mechanical, physical, electrical, and electronic properties. Therefore, engineering materials can be classified into five groups based on their nature as shown in Fig. 1.2.

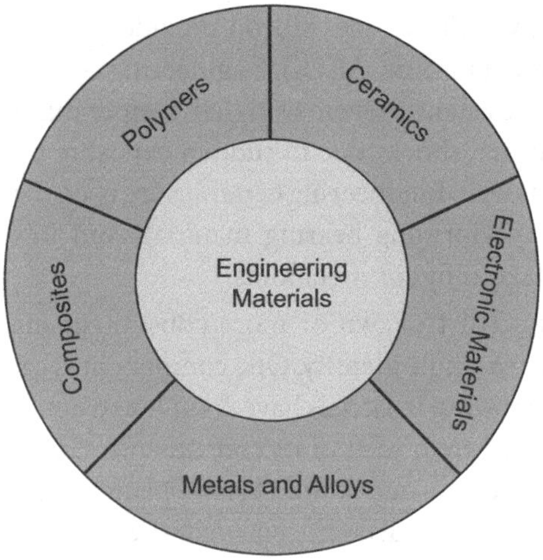

Fig. 1.2: Five major groups of engineering materials

Metals and alloys are referred to as metallic materials. Metallic materials are composed of one or more metallic elements (e.g., iron, copper, aluminum, titanium, gold, platinum), and often non-metallic elements also (e.g., carbon, oxygen, nitrogen) in relatively small amounts. Metallic materials have a crystalline structure in which the atoms are arranged in an orderly manner and are relatively dense as compared to ceramics and polymers. They are relatively strong and ductile at room temperature, and many maintain good strength even at high temperatures. Metals and alloys in general are good thermal and electrical conductors due to the presence of non-localized electrons. Due to the significantly higher production and uses of iron-based alloys as compared to other alloys, metallic materials are commonly divided into two classes: (a) *ferrous metals and alloys (iron and steels)* and (b) *nonferrous metals and alloys* (aluminum, copper, lead, zinc, titanium, nickel, and cobalt). The types of ferrous and nonferrous alloys and their applications will be discussed in detail in Unit 3.

Polymers consist of long molecular chains or network structures that are usually based on carbon-containing organic systems and are popularly known as plastics as well. Most of the polymers are non-crystalline in nature. Some polymers consist of mixtures of crystalline and non-crystalline states. Polymeric materials are relatively lighter and have relatively low softening temperatures as compared to metals and alloys. Industries are developing polymer-polymer mixtures which are known as *alloys* or *blend* to achieve the required properties. Plastic materials have been the fastest-growing basic material. Therefore, this type of materials continues to replace metals, glass, and paper in automobile, aerospace, construction, and household materials, for which plastics are suitable. Polymer market is a growing area because of their low cost and suitable properties. Commonly, polymers are poor conductors of electricity due to the nature of their internal structure. However, conducting polymers are also available nowadays.

Ceramic materials consist of metallic and non-metallic elements which are chemically bonded together. They are inorganic materials and mostly are oxides, carbides, and nitrides. These materials can be crystalline, non-crystalline, or mixtures of both. Refractories, glass, clay products, and cement are *traditional ceramics.* Ceramic materials have high hardness, high- temperature strength, good wear and heat resistance and are mostly brittle in nature. The thermally insulative properties, along with the high heat and wear resistance of many conventional ceramics, make them useful for furnace linings for melting of metals/alloys and heat treatment. *Advanced engineering (engineering or structural) ceramics* have improved properties, which includes silicon dioxide (SiO_2), silicon carbide (SiC), silicon nitride (Si_3N_4) and aluminum oxide (alumina, Al_2O_3). Engineering ceramics have improved fracture resistance, better wear and corrosion resistance (even at higher temperatures and in adverse environments), and enhanced resistance to thermal shock (due to sudden exposure to very high or very low temperatures) and many are light weight as well. Engineering ceramics are used for advanced applications in automobile engines, jet engines, high performing bearing materials and in electronic industries that have been normally dedicated to metals, alloys or metalloids.

Composite materials consist of two or more constituent materials (metal, ceramic or polymer) physically integrated to form a single identity. One component is called matrix and another is called filler or reinforcing material. Composite materials have complex structures. The final properties of composites will be different and enhanced than each of its constituents. Composite materials can be classified into natural and synthetic. Wood, bones, teeth, mud bricks, plant leaves, bird feathers are examples of natural composites. The application of natural composite materials is well documented in our ancient literature. Synthetic composites are based on metal/polymer/ceramic matrix. For example, the matrix material may be a metal such as copper/aluminum, or a polymer such as epoxy/polyaniline, or a ceramic such as Silicon carbide/alumina. Depending on the type of matrix used, the composite may be classified as a *metal matrix composite (MMC)*, a *polymer matrix composite (PMC)*, or a *ceramic matrix composite (CMC)*. Composites can be of many types depending on the selection of matrix and reinforcements. Fibrous and particulate reinforced composites are mainly used in applications. Commonly, the components do not dissolve in each other, and they can be physically identified by an interface between them. Therefore, interface characteristic is a very important parameter in the development of composites. Composite materials have replaced many metallic components due to their lower density, higher strength, and better stiffness-to-weight ratio. Aluminium (Al) matrix reinforced with silicon carbide (SiC) is a typical example of the MMC used in automotive applications. The application of composite materials is growing, especially in aerospace, avionics, automobile, civil structural, and sports equipment industries.

Electronic materials are extremely important for advanced engineering technology. Electronic materials can be a single crystal or a polycrystalline system. The most important electronic material is pure silicon that is modified in various ways to tailor its electrical characteristics to suit the intended application. *Electronic materials* such as Silicon, Graphene, Gallium Nitride (GaN), etc. have multifunctional properties. The application of electronic materials in the semiconductor industry is enormous and the market of electronic materials is growing. Examples of electronic materials: computer, mobile phone, electronic display panel, integrated circuits (IC), microelectronic devices, supercomputer, robot, etc. With the advances of nanotechnology, electronic materials will undoubtedly play a vital role in the "factories of the future".

1.3.2 Classification of Engineering Materials

Engineering materials can also be classified by merit of their usage in three major groups namely (i) structures, (ii) machines, and (iii) devices.

Structures refer to the items that are erected by engineers. Materials used in melting and heat treatment furnaces, road, oil refinery tower, bridges, buildings etc. are examples of structural materials.

Machines refer to moving objects such as jet engines, lathe, electric motors, automobiles and generators and all materials used in machines fall under this category.

Devices are the recent addition to engineering materials and refer to innovations in electrical, semiconductor and electronics industries such as shape memory alloys, sensors, multifunctional materials, a photoelectric cell, piezoelectric pressure gauges, etc.

1.4 SPACE LATTICE, CRYSTAL AND UNIT CELLS

To understand the geometry of crystals, we need to look into the framework with reference to which a crystal structure can be described. Three-dimensional infinite array of points is known as *Space Lattice* in which every point has identical surroundings to that of every other point in an array. Each point of the space lattice is known as the lattice *Point or Lattice site*. An array of entities (atoms, ions or molecules) in space, having at least translational symmetry (in simple words the pattern looks identical at all locations in the lattice) is called crystal, which is unique for each crystalline material. The said entity is known as the basis or motif. A motif is a group of atoms/ions associated with the lattice points and organized in a certain arrangement relative to each other. It is also noted that a geometrical entity or a physical property or a combination of both can serve as a motif. The lattice points are the assigned places for motifs/basis in a crystal. When each lattice point will be occupied by an entity (atom/ion/molecule) then the space lattice will be converted to *crystal*. Therefore, we can say that *Crystal = Lattice + Motif (or Basis)*.

In an ideal crystal the grouping of lattice points about any point is identical with the grouping of lattice points about any other lattice points in the crystal lattice. Unit cell is the smallest unit of a space lattice. We can create a space lattice by repeating unit cells. It is important to note that all atoms do not necessarily coincide with their respective lattice points.

Now, we should learn how to generate space lattices. Just to start, we have to visualize the periodic arrangement of points on a sheet of paper. Fig. 1.3 shows the 2D arrangement of points on the paper.

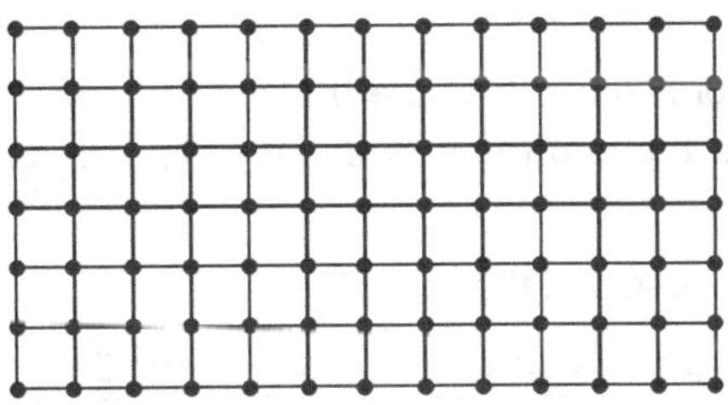

Fig. 1.3: A two-dimensional array of points on the plane of paper

This is a two-dimensional square array of points in space. Two non-coplanar vectors are required to represent the two-dimensional square lattice. Fig. 1.4 shows 1D and various types of 2D lattices

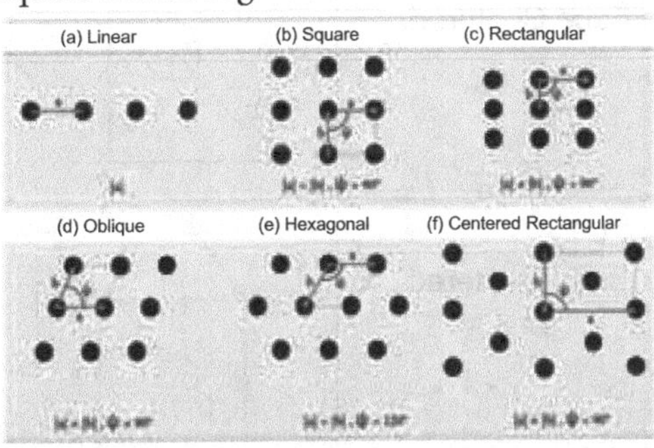

Fig. 1.4: Typical illustration of 1D and five 2D lattices

Space lattice is generated by repeated translation of three non-coplanar vectors namely **a**, **b**, and **c**. The characteristics of three non-coplanar vectors originating from one corner of the unit cell will decide the size and shape of the unit cell. Fig. 1.5 shows a portion of crystal lattice and parameters of a unit cell.

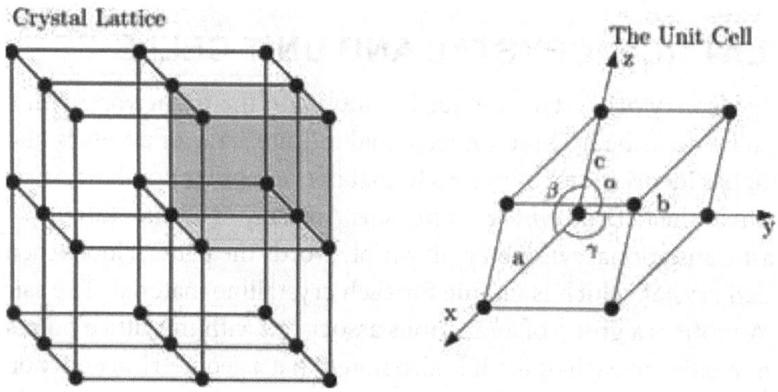

Fig. 1.5: A portion of a three-dimensional cubic lattice and illustration of the parameters of a unit cell.

There are only 14 different possible ways in which points can be arranged in 3D space such that each point has identical surroundings. These are called Bravais Lattices. Every lattice point in Bravais lattice represents one constituent particle which may be an atom, an ion or a molecule. Lattice points in a crystal lattice are joined by straight lines which decide the geometry of the lattice. The axial lengths (length of unit cell edges) a, b, and c and the interaxial angles α (angle between **b** and **c**), β (angle between **a** and **c**), and γ (angle between **a** and **b**) are called as lattice constants of a unit cell as shown in Table 1.4. It is important to note that these six parameters, a, b, c, α, β, and γ are required to characterise a unit cell.

The crystal has inherent symmetry. The 14 Bravais Lattices can further be classified to 7 types of crystal system purely based on symmetry. This classification is not based on the geometry of the unit cell. Example: if a crystal has only one 4-fold axis then it would be classified as a tetragonal crystal. Table 1.5 provides the details about the minimum symmetry that is needed to be present in case of various crystal systems.

Table 1.4: 7 Crystal systems and their possible variations along with unit cell with examples

Crystal system	Space lattice	Unit Cell	Typical Examples
1. Cubic $a = b = c$ $\alpha = \beta = \gamma = 90°$ 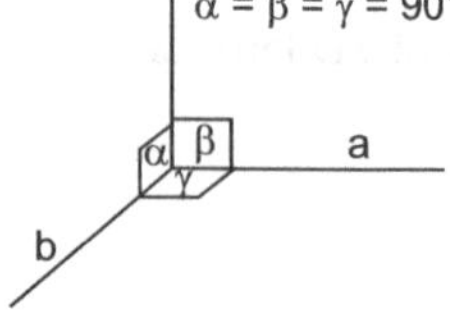	Primitive	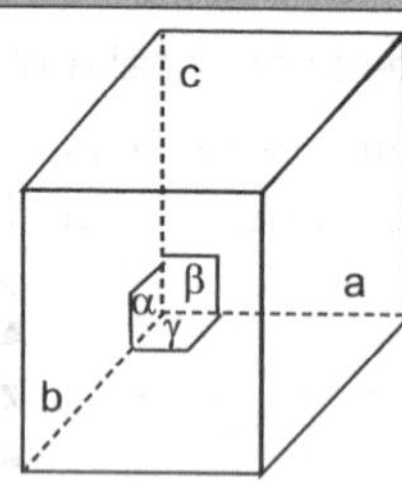	Polonium (α – Po)
	Body Centered		Cr, W, Mo, Nb, α–Fe, β–Ti, V
	Face Centered		Al, Cu, Ag, Au, Pt, Ni, Pb, Pd, Rh, Ca
2. Tetragonal $a = b \neq c$ $\alpha = \beta = \gamma = 90°$ 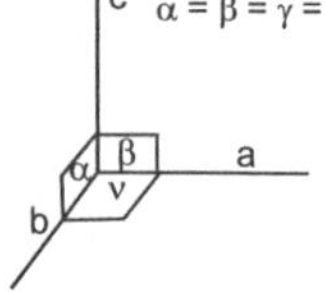	Primitive	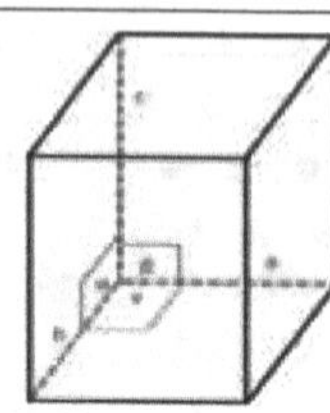	None in pure form
	Body Centered		Tin, Indium, Protactinium

3. Orthorhombic 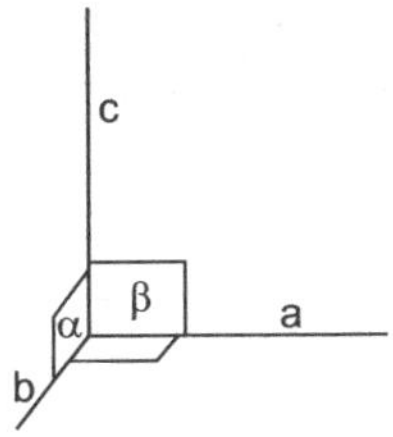	Primitive	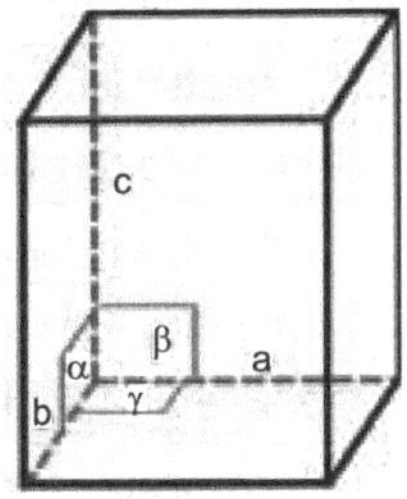	None in pure form (Fe_3C, CdTe)
	Body Centered		None in pure form (SiS_2, GaAs)
	Face Centered		None in pure form (α-S)
	End Centered		None in pure form (α-U)
4. Rhombohedral 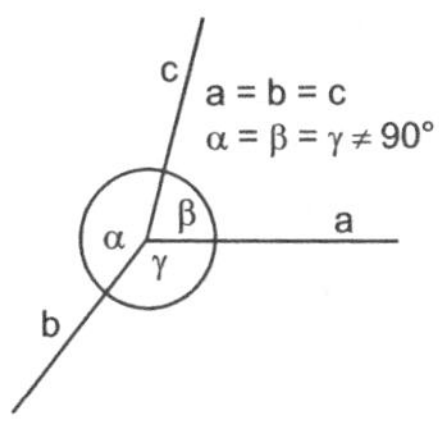	Primitive	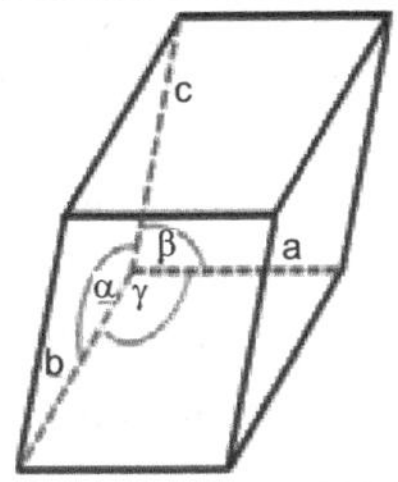	Hg, Bi, Po, Pa
5. *Hexagonal* 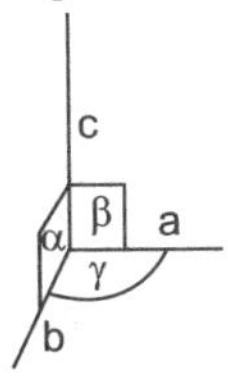	*Primitive*	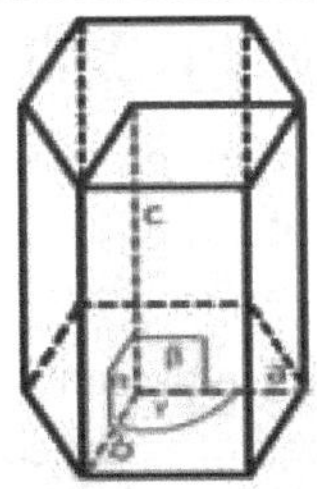	Mg, Cd, Zn, graphite, CdS, ZnO
6. Monoclinic 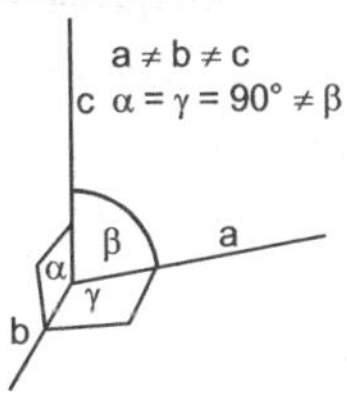	Primitive	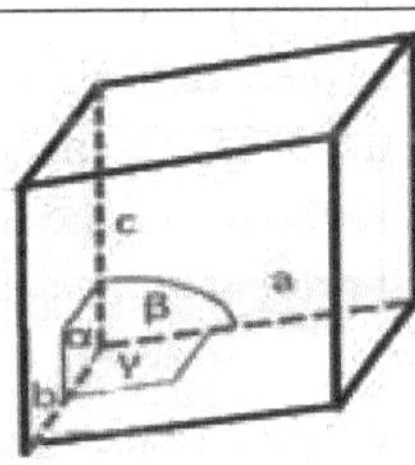	Hydrated sodium sulphate ($Na_2SO_4.10H_2O$) Monoclinic sulphur
	End Centered		
7. Triclinic 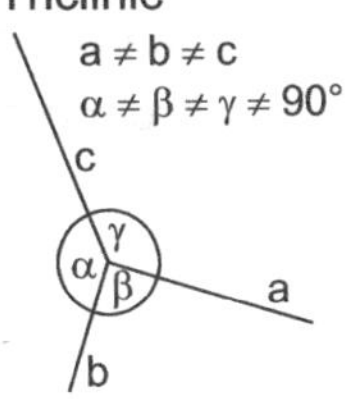	Primitive	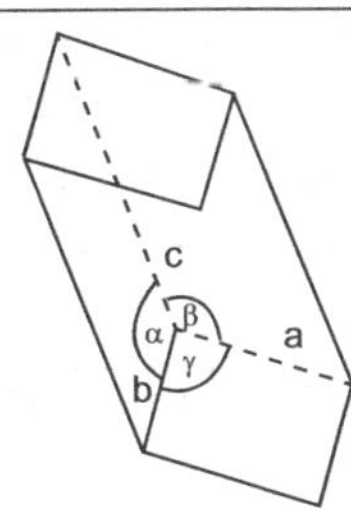	Hydrated copper sulphate ($CuSO_4.5H_2O$), $K_2Cr_2O_7$, H_3BO_3

The order of symmetry present in 7 types of crystal system can be represented in decreasing order of symmetry as cubic, hexagonal, tetragonal, trigonal, orthorhombic, monoclinic, and triclinic. The cubic system has the highest order of symmetry, and the triclinic system has no symmetry. It is worthwhile to understand the meaning of characteristic symmetry. It is mentioned in Table 1.5 that a cubic system has four 3-fold rotation characteristic symmetry.

Table 1.5: Seven crystal systems and their associated characteristic symmetry

Crystal system	Characteristic symmetry
Cubic	Four 3-fold rotation axes (two will generate other two)
Hexagonal	Only one 6-fold rotation axis
Tetragonal	Only one 4-fold rotation axis
Trigonal	Only one 3-fold rotation axis
Orthorhombic	Only three 2-fold rotation axis
Monoclinic	Only one 2-fold rotation axis
Triclinic	None

The rotational, translational and reflection symmetry operations constitute (3-fold rotation symmetry) the symmetry elements of a crystal. Four-fold symmetry means that during each complete rotation about an axis (passing through the centres of two opposite faces of the unit cell i.e. cube), the crystal will acquire identical position in space four times.

1.5 PRIMITIVE AND NON-PRIMITIVE UNIT CELLS

Unit cells can be classified into two types, *primitive* and *non-primitive unit cells.* When motifs/basis are present only on the corner position of a unit cell, it is called a primitive *unit cell.* There are seven types of primitive unit cells.

When one or more motifs/basis are present at positions other than corner position in addition to those at corner positions of a unit cell, it is called a non-primitive *unit cell* or *centered unit cell.* Non-primitive unit cells are of three types: body-centered unit cells, face-centered unit cells and end-centered unit cells. *Body-centered unit cells or body centered cells* contain one motif/basis (atom/ion/molecule) at its body-centered position besides the ones present at its corners. *Face-centered unit cells* or face-centered cells contain one motif present at each face of the unit cell besides the ones present at its corner. *End-centered unit cells* contain one motif present at the centre of any two opposite faces of the unit cell besides the ones present at its corners.

Lattice positions, directions and planes are important terms in case of Bravais lattices. In fact, 14 Bravais lattices are the skeletons for a large number of crystalline solids. There are standard methods for describing the geometry of crystalline structures (planes and direction). William Hallowes Millar (British crystallographer) has proposed a few basic rules to describe the geometry in and around a unit cell. This description is beyond the scope of this book.

1.6 CUBIC AND NON-CUBIC STRUCTURES

The cubic structure is the main crystal structure associated with important engineering materials. The crystal structures of a few metals are listed in Table 1.6. As we see from an observation of Table 1.6, most metals at room temperature are found in either one of the body centered cubic (BCC), face-centered cubic (FCC), body-centered tetragonal (BCT), or hexagonal close-packed (HCP) structure.

The hard ball or sphere model is commonly used to describe and understand the characteristics of unit cells. Here, motifs or basis is indicated by hard balls or spheres. The sphere is representative of each motif i.e. atom in the case of a metallic system. The radius of the sphere is the atomic radius.

Atomic radius is the total distance from the centre of the nucleus of an atom to the outermost orbital of its electron. The position of the electron is not fixed. Thus, the *atomic radius* is the mean distance from the centre of the nucleus to the boundary of the surrounding shells of outermost electrons. However, it is very difficult to determine the accurate atomic diameter due to the uncertainty in the position of outermost electrons (Reference: Heisenberg's uncertainty principle).

A reduced-sphere unit cell model is commonly used for the sake of simplicity and to provide a better illustration to the learners. The reduced-sphere unit cell does not depict the actual size of the motif and hence only represents the centre of the motif occupying that position. Thus, the sphere or hard-ball model is used to determine the coordination number, packing efficiency, and void positioning in the unit cell which direct the various properties of the crystalline materials. In many elemental crystals, the motif will be one atom per lattice point.

These crystals are known as monoatomic crystals. The term monoatomic is often not used. However, a FCC crystal means a monoatomic FCC crystal, unless otherwise stated. It is true in the case of other crystal systems also.

Table 1.6: Crystal structure details of few elemental solids

Atomic Number	Pure Metal	Structure	Unit cell dimensions (nm)	
			a	b/c
4	Beryllium	HCP	0.229	0.358
12	Magnesium	HCP	0.321	0.521
13	Aluminium	FCC	0.405	-
22	Titanium	HCP	0.295	0.468
23	Vanadium	BCC	0.303	-
24	Chromium	BCC	0.289	-
25	Manganese	Cubic	0.891	-
26	Iron	BCC	0.287	-
27	Cobalt	HCP	0.251	0.409
28	Nickel	FCC	0.352	-
29	Copper	FCC	0.362	-
30	Zinc	HCP	0.267	0.495
39	Yttrium	HCP	0.365	0.573
40	Zirconium	HCP	0.323	0.515
41	Niobium	BCC	0.330	-
42	Molybdenum	BCC	0.315	-
45	Rhodium	FCC	0.380	-
46	Palladium	FCC	0.389	-
47	Silver	FCC	0.409	-
48	Cadmium	HCP	0.298	0.358
57	Lanthanum	HCP	0.376	0.606
72	Hafnium	HCP	0.320	0.506

73	Tantalum	BCC	0.331	-
74	Tungsten	BCC	0.317	-
77	Iridium	FCC	0.384	-
78	Platinum	FCC	0.392	-
79	Gold	FCC	0.408	-
81	Thallium	HCP	0.346	0.553
82	Lead	FCC	0.495	-

1.6.1 Primitive (Simple) Cubic Unit Cell

In Fig. 1.6, a unit cell of monoatomic primitive (simple) cubic unit cells have been depicted in two ways using a hard-ball (sphere) model and a reduced-sphere model. In the case of primitive cubic crystals, the number of nearest neighbouring atoms is six (6) as shown in Fig. 1.5. Polonium (α–Po) crystallizes in the primitive cubic system. Thus, the coordination number of polonium atoms in polonium (α–Po) is six.

Primitive cubic unit cell has atoms only at its corner. At room temperature, there is a polonium atom at each corner of the unit cell. Each polonium atom at a corner is shared between eight adjacent unit cells as shown in Fig. 1.6.

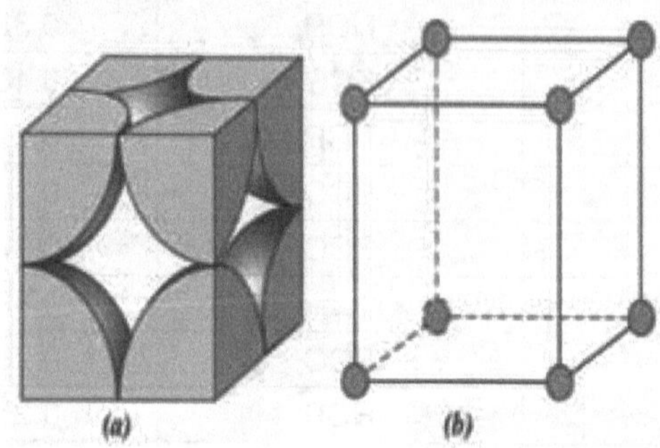

Fig. 1.6: Primitive cubic crystal structure, (a) a hard-ball (sphere) unit cell, and (b) a reduced-sphere unit cell.

It is clear that each atom in a simple cubic system is shared by four-unit cells in the same layer and four-unit cells of the lower (or upper) layer. Therefore, only 1/8th of a motif or basis (atom/ion/molecule) belongs to a unit cell under consideration as shown in Fig. 1.6. Since each primitive cubic unit cell has 8 atoms on its corners, therefore, the total number of atoms in one unit cell is $8 \times \frac{1}{8} = 1$ atom.

The effective number of atoms per unit cell of α–Po is only 1 (one) atom.

The materials with a primitive cubic structure have atoms contact along the edge of the unit cell. Thus, the side or edge length of a cube 'a' and the radius of each atom 'r' are correlated as , where 'a' is a lattice parameter.

The volume of the cubic unit cell $= a^3 = (2r)^3 = 8r^3$

There is one atom per unit cell in a primitive unit cell and we are assuming that all atoms are identical and spherical.

The volume of occupied space by atoms (number of effective atoms per unit cell × volume of single atom) $= 1\times\frac{4}{3}r^3 = \frac{4}{3}r^3$

The *packing fraction* of a unit cell is the fraction of volume of a unit cell occupied by the atoms. The *packing efficiency* of a unit cell is the percentage of volume of unit cell occupied by the atoms.

Thus, the *packing efficiency* of primitive cubic cell is

$$\frac{\text{volume of one atom}}{\text{volume of cubic unit cell}}\times 100\%$$

$$= \frac{\frac{4}{3}r^3}{8r^3}\times 100\% = \frac{1}{6}\times 100\% = 52.4\%$$

The packing factor of a primitive cubic unit cell is 0.52.

1.6.2 Body Centered Cubic (BCC) Unit Cell

In Fig. 1.7, a unit cell of monoatomic body-centered cubic (BCC) unit cells has been depicted in two ways using a hard-ball (sphere) model and a reduced-sphere model. This is the BCC Bravais lattice, where the number of nearest neighbouring atoms is 8 (eight) as shown in Fig. 1.7. There is one atom at the $\frac{1}{2},\frac{1}{2},\frac{1}{2}$ position considering the origin at any corner of the unit cell and one-eighth atom at each of the eight-unit cell corners. Each BCC unit cell has 8 atoms on its corner and 1 atom at its centre. Each corner atom is shared by eight-unit cells. Therefore, the total number of atoms in one unit cell is $\left(8\times\frac{1}{8}\right)+1 = 2$ atom thus making the effective number of atoms per unit cell of nickel at room temperature as 2 (two) atoms in the BCC unit cell. There are two atoms per unit cell in a BCC unit cell and we are assuming that all atoms are identical and spherical.

In the case of BCC crystal, the number of nearest neighbouring atoms is 8 (eight) as shown in Fig. 1.7. Iron crystallizes in the body-centered cubic system which is known as α–iron (α–Fe). The effective number of atoms per unit cell of iron at room temperature (α–Fe) is 2 (two) atoms. At room temperature, every iron atom is surrounded by eight adjacent iron atoms, whether the atom is located at the center or at the corner of the unit cell. Thus, the coordination number of iron atoms in α–Fe is eight.

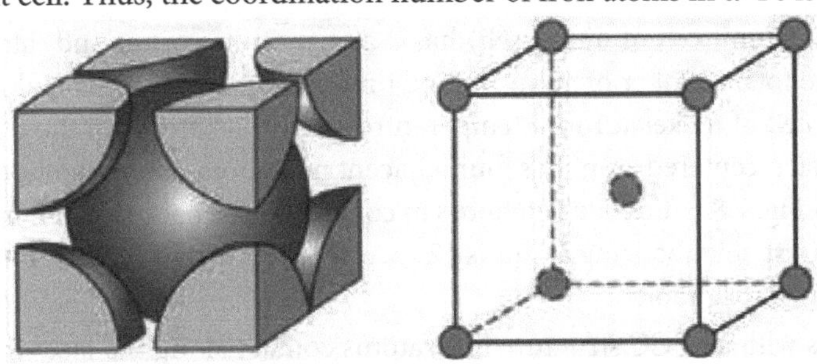

Fig. 1.7: Body-centered cubic crystal structure, (a) a hard-ball (sphere) unit cell, and (b) a reduced-sphere unit cell.

The materials with a BCC structure have atoms contact along the diagonal of the unit cell as shown in Fig. 1.7. Thus, the diagonal of the cube 'd' (BCC unit cell) and the radius of each atom 'r' are correlated as $d = 4r$ and $d = \sqrt{3}a$, where 'a' is a lattice parameter. Thus, $4r = \sqrt{3}a$

The volume of the cubic unit cell $a^3 = \left(\frac{4}{\sqrt{3}}r\right)^3$

Thus, the *packing efficiency* of BCC unit cell is

$$\frac{\text{volume occupied by effective number of atoms in BCC}}{\text{volume of cubic unit cell}} \times 100\%$$

$$= \frac{2\times\frac{4}{3}r^3}{\left(\frac{4}{\sqrt{3}}r\right)^3} \times 100\% = 68\%.$$

1.6.3 Face Centered Cubic (FCC) Unit Cell

In Fig. 1.8, a unit cell of monoatomic face-centered cubic (FCC) unit cells has been depicted in two ways using a hard-ball (sphere) model and a reduced-sphere model. There is one-half atom in the centre of each unit cell face and one-eighth atom at each of the eight-unit cell corners. Nickel (Ni) crystallizes in a face-centered cubic system.

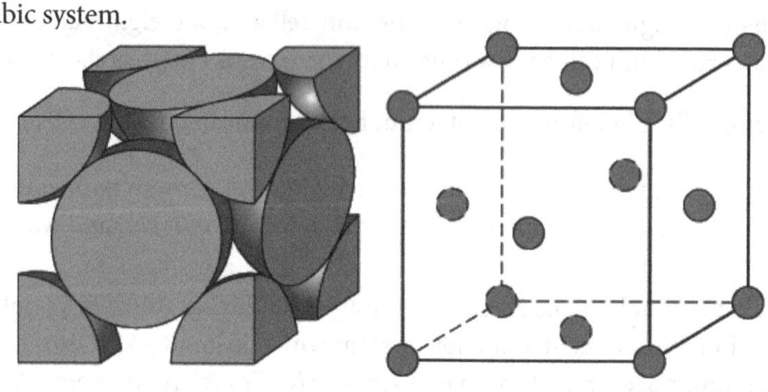

Fig. 1.8: Face-centered cubic crystal structure, (a) a hard-ball (sphere) unit cell, and (b) a reduced-sphere unit cell.

Since each FCC unit cell of nickel (Ni) has 8 atoms on its corner and atom at 6 faces of the unit cell. Therefore, the total number of atoms in one unit cell is (atoms, thus making the effective number of atoms per unit cell of nickel at room temperature as 4 (four) atoms in the FCC unit cell. In the FCC system, the front face-centered atom has four adjacent neighbors, four neighbors in contact on the front side (not visible in Fig. 1.8), and four neighbors in contact on the back side. Thus, in case of FCC crystals, the number of nearest neighbouring atoms is 12 (twelve) as shown in Fig. 1.8. It means each atom in FCC crystal has 12 nearest neighbors.

The materials with an FCC structure have atoms contact along the face diagonal of the unit cell as shown in Fig. 1.8. Thus, the face diagonal of a cube 'd_f' and the radius of each atom 'r' are correlated as $d_f = 4r$, and $d_f = \sqrt{2}a$, where 'a' is a lattice parameter. Thus, $4r = \sqrt{2}a$.

The volume of the cubic unit cell = $a^3 = \left(\frac{4}{\sqrt{2}}r\right)^3$.

Thus, the *packing efficiency* of BCC unit cell is

$$\frac{\text{Volume occupied by effective number of atoms in FCC}}{\text{volume of cubic unit cell}} \times 100\%$$

$$= \frac{4 \times \frac{4}{3}r^3}{\left(\frac{4}{\sqrt{2}}r\right)^3} \times 100\% = 70\%.$$

The packing efficiency of a FCC metal (74%) is greater than the packing efficiency of a BCC (68%) metal. This difference is to be expected, since each atom in FCC metal has 12 nearest neighbours (i.e. coordination number) whereas BCC metal has only 8 nearest neighbours. In the case of the FCC system, there is a specific type of atomic pattern. This pattern of layers is often indicated as ABCABC……type of stacking of atoms. In fact, the packing efficiency of 74% is the highest value possible for filling space by stacking equal-sized hard spheres. Thus, this structure is also called cubic close packing (CCP). The FCC and CCP are the same in the cubic system.

In whatever way motifs (atoms/ions/molecules) are packed, there is always some free space left in the unit cell. The free space present in the unit cell is designated as void. There are two types of voids namely tetrahedral and octahedral voids. The tetrahedral void is surrounded by four atoms whereas the octahedral void is surrounded by six atoms. The number of tetrahedral voids and octahedral voids in a unit cell of FCC or CCP lattice are 8 and 4 respectively. The technological importance of these voids will be discussed in Unit 2.

Apart from metals, there are several systems which have FCC structure. The diamond cubic (DC) structure is also FCC with two atoms per unit lattice point which is a repeating pattern of 8 atoms per unit cell. The motif contains two atoms of carbon, one at the lattice point and another one along the body diagonal (not a lattice point). The packing efficiency of DC is 34% only. Silicon has a diamond cubic structure. The NaCl structure is also FCC. In NaCl, motif contains both and ions. The molecular structure of solid methane is also FCC with five atoms (four hydrogen and one carbon) as a motif occupying the FCC lattice positions.

1.6.4 Non-Cubic Unit Cell

There are several non-cubic Bravais lattices as mentioned in Table 1.4. We will discuss a more common non-cubic system among metallic materials i.e. hexagonal crystal system. Fig. 1.9 represents hexagonal structure using a reduced-sphere model. HCP structure is characterized by its typical stacking arrangement of atoms. Each atom in one layer is located directly above or below the interstices among three atoms in the adjacent layers. This means that the tetrahedral voids of the second layer are covered by the atoms of the third layer. Thus, the atoms of the third layer and first layer are exactly aligned. This pattern is written as ABABAB….. type of stacking pattern. This hexagonal structure is called the hexagonal close-packed (HCP) structure.

The FCC structure has ABCABC……. type of atomic layer sequence whereas HCP structure has

ABABAB......type of atomic layer sequence. This subtle difference in stacking sequence severely alters the mechanical behaviour of HCP metals with respect to FCC metals. Several metals such as Beryllium, Magnesium, Zinc, Cadmium, Cobalt and Zirconium are examples of HCP systems.

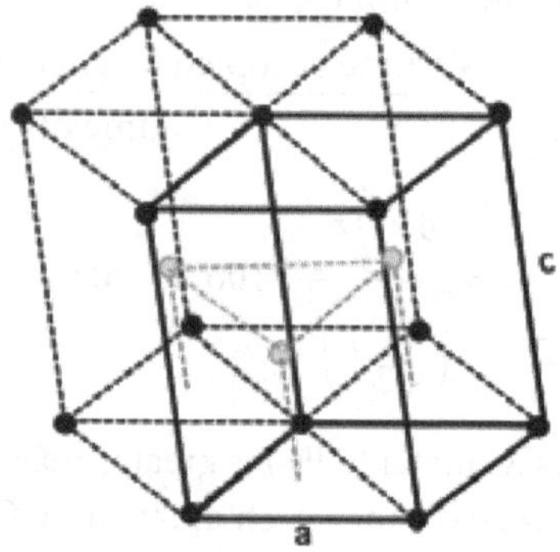

Fig. 1.9: Schematic showing hexagonal close-packed structure

1.6.5 Miller Indices

The properties of crystalline material such as metals and alloys vary with crystallographic orientation (including information about the direction and planes). In crystalline material, there are specific crystallographic directions and planes which contain a large number of atoms/ions. Thus, it is essential to refer to a widely used convention to specify the directions and planes in a crystal which is known as Miller indices (in the honour of English crystallographer William Hallowes Miller) of directions/planes.

1.6.5.1 Miller indices of direction in cubic crystal

A crystallographic direction is defined as a line vector directed between two points. Many properties of crystalline material vary with crystallographic direction. For a cubic crystal, the Miller indices are the vector component of the direction resolved along each of the coordinate axes and reduced to the smallest integer. In crystallography the positive y axis is the direction to the right of the paper, the positive z axis is direction to the top, and the positive x axis is the direction coming out of the paper.

The procedure for determining the Miller indices of direction in a cubic crystal is as follows:

1. Use rectangular x, y, z axes (right-handed) to locate atomic position in cubic crystal.
2. Identify the start and end points of a typical direction.
3. Position the vector, such that start (S: (x_1, y_1, z_1)) and end points (E: (x_2, y_2, z_2)) are lattice points and note the value of the coordinates.
4. Subtract to obtain: $((x_2-x_1), (y_2-y_1), (z_2-z_1))$.
5. Write these numbers $(x_2-x_1), (y_2-y_1), (z_2-z_1)$ in square brackets without the 'comma': after making them the set of least integers (here either multiply or divide these numbers by a common rational number).

   Note: In case of negative coefficient of Miller indices, we designate them with an over-bar i.e., $\overline{N}$.

The length of the vector represented by the Miller indices is $\sqrt{(X_2-X_1)^2+(Y_2-Y_1)^2+(Z_2-Z_1)^2}$.

Example 1.1: Determine the Miller indices of direction passing from $\left(0,0,\frac{1}{2}\right)$ to $\left(0,\frac{1}{2},0\right)$. Calculate the length of this vector.

Solution of example 1.1: Here, x_1, y_1, and z_1 are 0, 0, and $\frac{1}{2}$ respectively. Similarly, x_2, y_2, and z_2 are 0, $\frac{1}{2}$, and 0 respectively. Thus, (x_2-x_1), (y_2-y_1), and (z_2-z_1) are 0, $\frac{1}{2}$, and $-\frac{1}{2}$ respectively. Thus, the miller indices of the said direction is $[01\bar{1}]$.

Further, the length of this vector is $= \sqrt{0+(1)^2+(-1)^2} = \sqrt{2}$.

The Miller indices of important directions are shown in Fig. 1.10. The unit cell edge ([001] type) and face diagonals ([011] type) are shown in Fig. 1.10 (a), whereas body diagonal ([111] type) is shown in Fig. 1.10 (b).

There may be another way to determine the Miller indices by assuming the initial point as an origin and then find out the coordinate of the end point by shifting the origin to the initial point without changing the rectangular x, y, z axes. Here, the initial point will be the new origin and the end point will change its coordinate due to shifting of origin from (0,0,0) to (x_1, y_1, z_1). Then, follow step 4 and 5 only (discussed in section 1.6.5) to determine the Miller indices of any direction in cubic crystal. Now let us try to determine the Miller indices of a direction by applying the said rule.

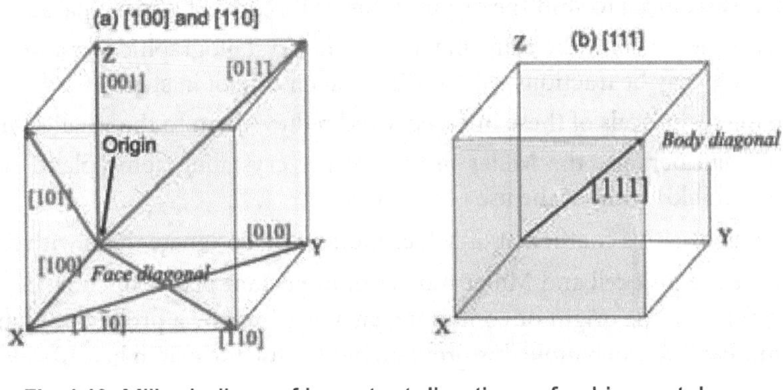

Fig. 1.10: Miller indices of important directions of cubic crystal

---

Example 1.2: Determine the Miller indices of direction passing from (0, 0, 1) to (0, 1, 0).

Solution of example 1.2: Initial point and final point are (0,0,1) to (0,1,0) respectively. Now, shift the origin to a point (0, 0, 1). Then, the coordinate of the end point in the new coordinate system will be (0, 1, −1). Thus, the Miller indices will be $[01\bar{1}]$.

A family of directions is a set of directions related by symmetry operations of the crystal. A family of directions is represented (Miller Index notation) as <u v w>. The angle Φ between two Miller indices $[xyz]$ and $[pqr]$ is given by the following relation

$$\cos\phi = \frac{xp + yq + zr}{\sqrt{x^2 + y^2 + z^2}\sqrt{p^2 + q^2 + r^2}}$$

The above formula is applicable only in case of orthonormal axes. Thus, the angle Φ between two Miller indices [xyz] and [pqr] for non-orthonormal axes will be different.

It is important to observe that parallel directions always have the same Miller indices. Further, it has been reported that deformation commonly occurs in the directions with the greatest linear density of atoms. The linear density is reciprocal of repetition distance in a particular crystallographic direction. The *repetition distance* is a vector between identical positions within a crystal. The repetition distance differs from structure to structure and from direction to direction. The repetition distances in the [110] direction is $\frac{a}{\sqrt{2}}$ and $a\sqrt{2}$ for FCC and BCC crystals respectively. The repetition distance for the family of directions is the same.

1.6.5.2 Miller indices of planes in cubic crystal

A crystalline material contains planes of atoms. Some specific planes are important for a crystal because these planes influence the properties and deformation behaviour of a crystalline material. Thus, it is essential to identify and understand various planes or groups of planes in a crystal. The Miller indices of a cubic crystal plane are defined as the reciprocal of the intercepts that the specific plane makes with three non-parallel edges of unit cell.

1. Use rectangular x, y, z axes (right-handed) to choose a plane that does not pass through the origin at (0, 0, 0). It is required to shift the origin in the unit cell itself if plane passes through origin.
2. Determine the intercepts of the plane in terms of the crystallographic x, y, and z axes for a unit cell. These intercepts may be fractions which will be taken care of in step 3.
3. Determine the reciprocals of these intercepts and reduce them to the smallest integer.
4. These whole numbers are the Miller indices of the crystallographic plane and are enclosed in parentheses i.e. (hkl) without the use of commas.

Note: In case of negative coefficient of Miller indices, we designate them with an over-bar i.e., $\overline{N}$.

Fig. 1.11 shows the unit cell and Miller indices of important planes of cubic crystal. Miller indices of planes passing through the origin or containing an axis would be a problem because their intercepts are not a finite number. We can handle this problem by shifting the origin because all the parallel planes possess the same indices.

A set of planes related by symmetry operations of the crystal system is called a family of planes. The family of planes is enclosed in { } brackets, while an individual member is enclosed in ( ) brackets. In cubic crystals the (hkl) plane is perpendicular to the [hkl] direction. The interplanar spacing ($d_{(hkl)}$) in cubic crystals is $\frac{a}{\sqrt{h^2 + k^2 + l^2}}$; where 'a' is the lattice parameter. The $d_{(hkl)}$ is the same for all members of a family of planes.

The determination of crystal structure is an important task for material scientists and engineers, which is carried out by X-ray Diffraction method. The X-rays of monochromatic wavelength () are reflected by a set of parallel planes in the crystal. The interplanar spacing of the crystal planes and angular position of the reinforced diffracted beam are correlated by using Bragg's law. The Bragg's law for the first order of diffraction is as follows:

$n\lambda = 2d_{(hkl)} \sin\theta [n = 1$, for first order diffraction]

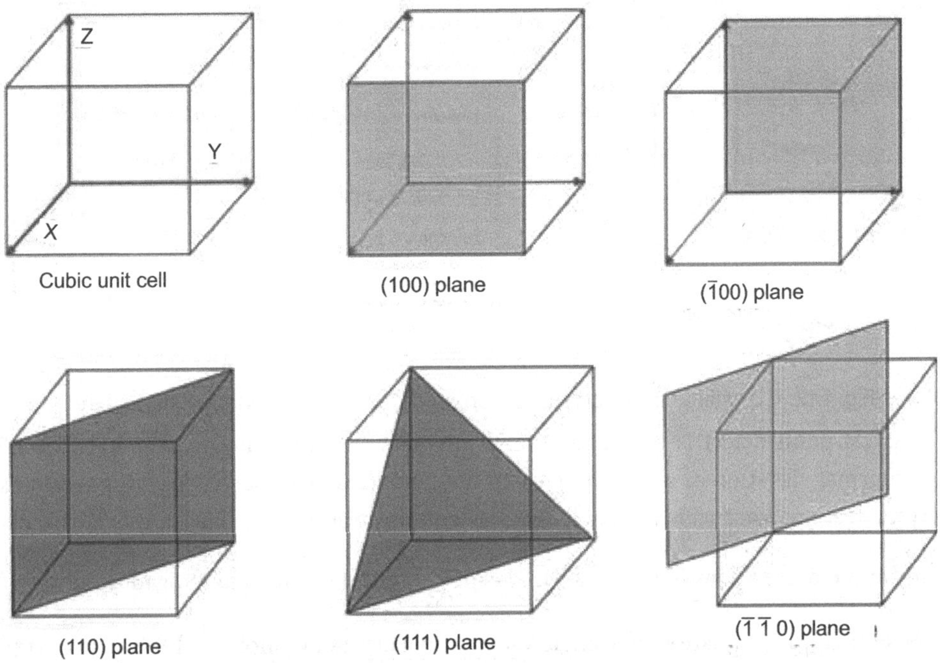

Fig. 1.11: Miller indices of planes in cubic crystal

Example 1.3: For a sample of iron crystal at 200C, calculate the lattice constant when diffraction from {110} planes was obtained at $2\theta = 44.7^0$. Assume the wavelength of incoming X-ray is 1.541 Å and there is the first order of diffraction.

Solution of example 1.3: In this case, $\theta = 22.35^0$. then sin 22.35 = 0.3803

Apply Bragg's law: $l = 2\, d_{(hkl)} \sin\theta$

$\frac{0.1541\text{nm}}{2*0.3803} = d_{(hkl)} = 0.2026$ nm. Thus, planar spacing of {110} planes is 0.2026 nm.

Iron is BCC at 20°C, therefore lattice constant and interplanar spacing will be correlated as relation $a = d_{(hkl)}\sqrt{h^2+k^2+l^2} = 0.2026 * \sqrt{2}$ [Note: Iron shows FCC (γ-iron) and BCC (δ-iron) crystal structure at higher temperatures. Hence, lattice parameter and interplanar spacing relation will be different].

Thus, the lattice constant of iron (a) is 0.287 nm.

1.6.6 Crystal Defects

Crystals are never perfect and always contain various types of defects. These defects affect the structure sensitive properties such as yield stress, tensile strength, ductility, fracture toughness, coercivity, electronic conductivity, cold formability, corrosion and wear resistance of metals and alloys. The structural imperfection of a crystal can be classified based on their dimensionality as point, line, surface/ interface, and volume defects. The classification of crystallographic defects is shown in **Fig. 1.12.** It is also true that most defects increase the free energy of the crystal.

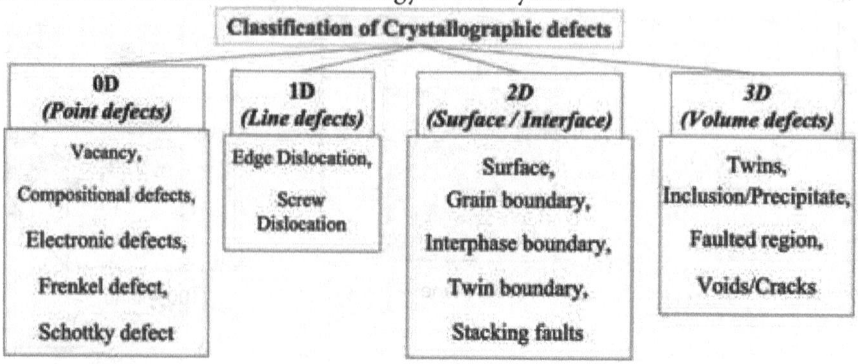

Fig. 1.12: Classification of crystallographic defects based on dimensionality

The simplest point defect is a vacancy, which involves an empty atomic site. Vacancy may arise either from thermal vibration of atoms at elevated temperatures or imperfect packing of atoms during original crystallization. Vacancies are considered as equilibrium defects. The number of vacancy in a metal at any temperature (T in Kelvin scale) is $N\,exp(\left(-\frac{\Delta H_v}{RT}\right)$ where N is Avogadro number (6.023 × 1023), is the enthalpy of formation of a single vacancy in a crystal(kJ/mol), and R is gas constant (8.314 J/mol.K).

The most common line defect in crystals is dislocation which plays an important role in strengthening of crystalline metals and alloys. There are two types of dislocations namely edge and screw dislocations. The edge dislocation may be described as an edge of an extra plane of atoms within a crystal. Screw dislocation may be described as a slip pattern within a crystal resembling the turning of a screw and is slightly difficult to visualize. Both type of dislocations are created during crystallization i.e. solidification of crystalline solids. They are more commonly formed by plastic deformation. Perfect edge or perfect screw dislocation are rare to be present in crystals. Most dislocations in crystals are of the mixed type, with edge and screw components.

External or internal discontinuities between grains or phases are described as boundaries and are known as *grain boundaries*. They are the surface defects in metals or alloys originated by crystals of different orientations meeting during crystallization of solid crystals. The structure of crystalline solid contains many crystals of different orientations. These individual crystals are called grains. Within a grain, all unit cells are arranged in one pattern and one orientation. The grain boundary is the boundary between two grains which is not aligned with any of these grains. The discontinuities between different phases are called the interphase boundary.

Other types of crystalline defects are mentioned in **Fig. 1.12**. Metallurgists and materials scientists control the amount and distribution of these defects to tailor the desired properties of crystalline solid materials.

1.7 BONDS IN SOLIDS

Basic points in this topic need to be revisited before starting the classification and characteristics of various types of chemical bonding. Atom is the basic unit of an element that can undergo chemical change. Atoms mainly consist of three basic subatomic particles i.e. proton, neutron, and electrons. The outer electrons (high energy electrons) are the valence electrons, and it is mainly their behavior that determines the reactivity of the atom. The electrons form a cloud of varying density around a denser atomic nucleus containing almost all the mass of the atom. Electrons obey the law of quantum mechanics, and as a result the energies of the electrons are quantized. An electron can have only certain allowed values of energies. When an electron changes its energy, then it can have only certain allowed values of energies. No two electrons in the same atom can have all four quantum numbers the same.

A ***chemical bond*** can be defined as an attractive force that holds groups of two or more atoms/ions/molecules together, causing them to behave as a single unit. Fig. 1.13 indicates the classification of chemical bonding. Let us try to understand why do atoms form bonds or how do atoms combine to form molecules. An element or a molecule will only be formed/stable if it has a lower free energy than the individual atoms. Atoms of the noble gases are inert and do not react with any other atoms and hence their molecules are monatomic (i.e. contain only one atom). The inertness of noble gas atoms is because the atoms have low energy, and it cannot be further reduced by forming a compound. The low energy state of noble gases is due to their complete outer shell electrons. Many molecules have this arrangement which is called a noble *gas structure.* The bonding of adjacent atoms is essentially an electronic process. Bonding in solids may be classified as either primary or secondary bonding as shown in Fig. 1.13.

Strong or Primary bonds (bond energy: 100–1000 KJ mol^{-1}) are formed when outermost shell electrons are transferred or shared between atoms. Weaker or secondary bonds (bond energy: 1–50 KJ mol^{-1}) are formed when there is slight attraction between positive and negative charges without actual transfer or sharing of electrons. Methods of primary bonding include the metallic, covalent, ionic or mixed bonds. In fact, secondary bonds are much weaker bonding mechanisms that are only predominant when one of the primary bonding mechanisms is absent.

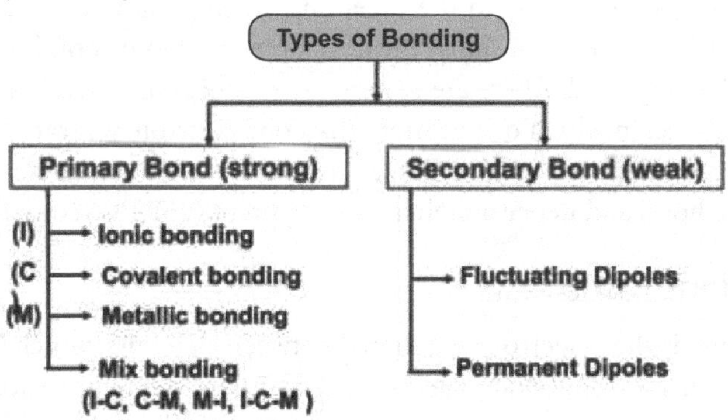

Fig. 1.13: Classification of chemical bonding.

1.7.1 The Ionic Bond

The *ionic bond* gets formed as a result of an electron transfer from one atom to another. There is *electrostatic* or *Coulombic* attraction between oppositely charged ions which is called an ionic bond.

In the ionic bond, the electrons are shared by opposite ions namely cation (electropositive ion) and anion (electronegative ion). Commonly, electron(s) in the outermost orbital shell can take part in ionic bonding. The cation gives up its valence electron(s), while the anion accepts them to complete orbitals or sub-orbitals of individual ions. Electropositive elements easily give up electrons and are typically Group I or II elements, e.g. Li, Na, K, and Ca, whereas electronegative elements easily take up electrons and are typically Group VI or VII elements, e.g. F, Cl, Br, N, and O. Ionic bonds between different atoms are preferred when the electronegativity difference between them is large.

Table salt (NaCl) is a common example of ionic compound. Let us understand the mechanism of formation of NaCl. We know that Na atom has one valence electron while Cl atom has seven valence electrons as shown in Fig. 1.14. Electropositive element easily gives up electrons and an electronegative element easily takes up electrons. As a result, the transfer of one electron from Na atom to Cl atom is favoured to achieve a more stable electron configuration (full outer orbital shells of eight electrons) for both ions. Here, Na is considered as cation due to the transfer of its electron and hence there is a net positive charge on Na^+ ion.

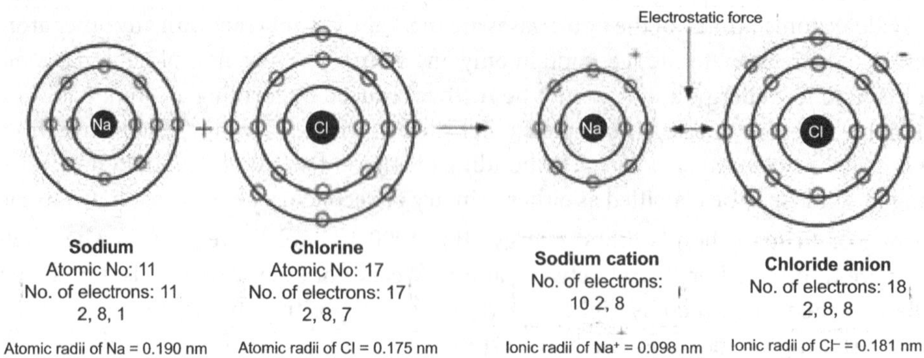

Fig. 1.14: Ionic bonding between sodium and chlorine atoms

The Cl is considered as anion due to the accepting an electron and hence there is a net negative charge associated with Cl^- ion. Fig. 1.14 shows the development of ionic bonds between Na and Cl atoms and formation of NaCl compound. There are several other ionic compounds such as NaBr, NaF, Na_2O, MgO, CaO, CrO_2, MoF_2 etc. in which one or more than one electron are transferred. In CaO, both ions are doubly ionised (i.e. two electrons are transferred from each calcium atom to an oxygen atom) leading to stronger interatomic bond and hence a higher melting point (2572°C) compared to NaCl (800°C).

1.7.2 The Covalent Bond

Covalent bond get formed when electrons are shared between elements, which leads to the formation of covalent compounds. The covalent bond enables the constituent elements to acquire their nearest noble gas configuration. The atoms will complete the outer electron shell by sharing with its neighbours. The sharing of electrons gives rise to covalent bonding. Covalent bond exists in case of carbon. The carbon has four electrons in its outer shell. Four more electrons are required to fill this shell and these may be acquired by sharing an electron with each of its four neighbours when carbon is in solid form. Fig. 1.15 describes the change in potential energy as a function of distance between two atoms. The lowest potential energy configuration will decide the equilibrium distance between two carbon atoms in solid form.

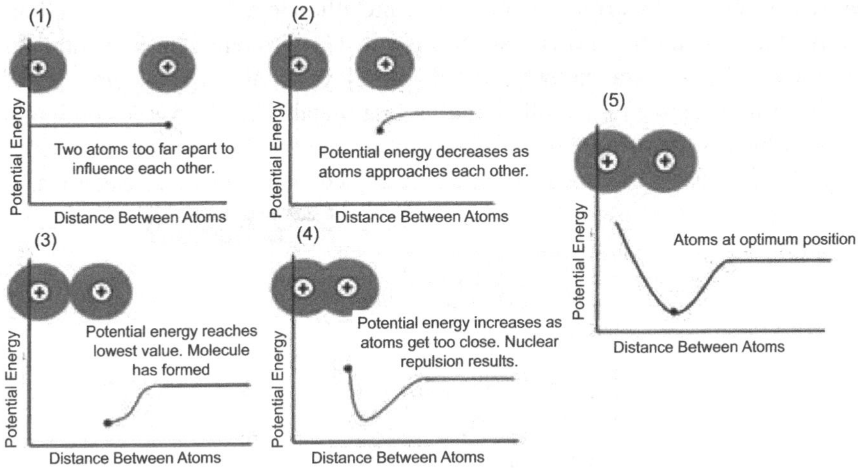

Fig. 1.15: Potential energy variation as a function of distance between two atoms

Oxygen molecule is a common example of covalent compound. This molecule is formed due to sharing of electrons between two oxygen atoms (shown in Fig. 1.16). Each oxygen atom is short of two electrons to form the nearest noble gas i.e. neon atom configuration. Therefore, when these atoms come closer and share two electrons with each other, they give rise to a double covalent bond between the two shared electron pairs of the atoms.

Covalent bonds are also possible for single and triple bonds where bonds are formed between one and three pairs of electrons respectively. The bond energy of a double bond is higher than a single bond provided both bonds are between similar types of atoms. The covalent bonds are directional in nature and covalent compounds are generally insoluble in water. Diamond, the hardest known naturally occurring substance on Earth, is formed from covalent bonds between carbon atoms arranged in a three-dimensional structure.

1.7.3 The Metallic Bond

Metallic bonds, as the name suggests, are bonds found only in metals. Metals are elements of the electropositive nature, thus it is very easy for the constituent atoms to lose their outer shell electrons and form ions. In metals, these positively charged ions are arranged in an orderly manner and held together in a sea of negatively charged free electrons. These free electrons are mobile and considered to be spread throughout the crystal.

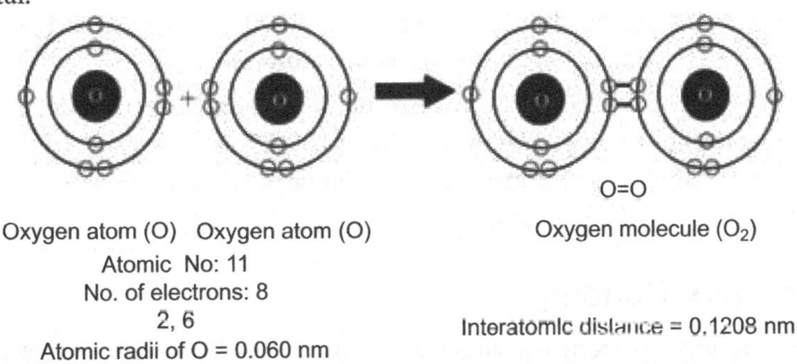

Fig. 1.16: Formation of covalent bond in oxygen molecule

However, it would be incorrect to think of a metallic element as an ion (cation) since the sea of electrons is shared by all the individual cations and hence quenching or neutralising the positive charge. The lustre and colour shown by typical metals/alloys are also due the presence of free electrons. Figure 1.17 shows an example of metallic bonding in a metal. The electronic configuration of sodium is $1s^2 2s^2 2p^6 3s^1$; which contains only one electron in its valence shell. In the solid-state, sodium metal contains an array of Na^+ ions that are surrounded by a sea of electrons i.e. 3s electrons.

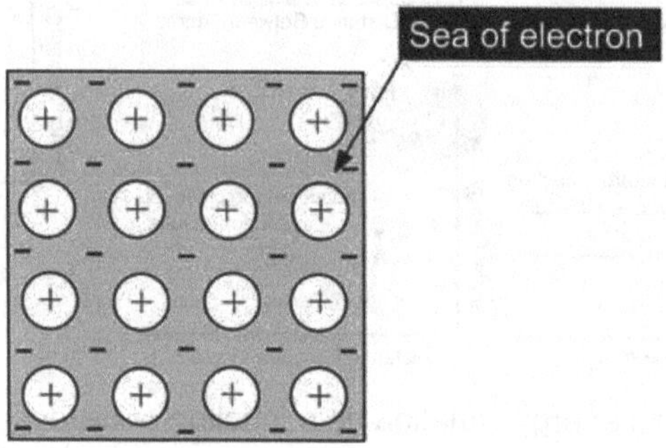

Fig. 1.17: A metallic crystal pictured as a lattice of +ve ions embedded in 'sea' of electrons.

In the case of an alloy, each metal atom contributes one or more valence electrons to create a pull of free or mobile electrons in the crystal. These mobile electrons are responsible for the high thermal and electric conductivities of metals and alloys. These mobile electrons are also responsible for the transfer of thermal energy from one end to another end of a metal/alloy.

1.7.4 Mixed Bonding

In compounds involving more than one element, ionic bonds are favoured when the electronegativity difference is large. In general electronegativity steadily increases across the Periodic Table from Group I to Group VIII. It also varies within a group, decreasing with increasing atomic number except in the case of transition elements. The transition from pure ionic to pure covalent is gradual. Calculations by Pauling illustrated that a 50-50 mixture occurs at an electronegative difference of 1.7. We can see this mixed bond in SiO_2. The electronegativities of Si and O are 1.9 and 3.44 respectively. The electronegativity difference is 1.54. We can see that in SiO_2 bonding, there is a 45% ionic character and a 55% covalent character.

A *polar covalent bond* is a bond that has a mix of ionic character and covalent character. A large difference in the electronegativity of two atoms indicates a greater ionic character and is considered a purely ionic bond; whereas a very small, negligible difference is considered a purely covalent bond. A polar covalent bond exists when the electronegativity difference is somewhere in between, generally more towards the covalent side (small electronegativity difference).

When a polar covalent bond is formed, the result is an unequal sharing of electrons between atoms. An example of a molecule with a polar covalent bond is hydrogen fluoride. In this molecule, hydrogen has a partial positive charge while fluorine has a partial negative charge.

Finally, Table 1.7 shows the typical properties of ionic, covalent and metallic bond.

1.7.5 Secondary Bonding

Secondary bonds are formed from the slight attractive forces between positive and negative charges. Secondary bonds, as opposed to primary bonds, are bonds with much smaller bonding energies that do not involve the transfer or sharing of electrons. These bonds are caused by permanent or temporary dipoles within the atom or molecule. They are weaker in nature and are broadly classified as Van der Waals forces and hydrogen bonds.

Table 1.7: Summary of properties with respect to ionic, covalent and ionic bonds

Property	Bonding Type		
	Ionic Bonding	**Covalent Bonding**	**Metallic Bonding**
Nature of involved atoms	Metal and Non-metal atoms	Non-metal atoms only	Metal atoms only
Structure details	Form giant lattices that are brittle so are easy to crush	Form molecules with weak forces of attraction so are mainly gases and liquids	Forms a strong lattice structure
Melting point (MP) or boiling point (BP)	High melting and boiling point {Sodium Chloride (NaCl): MP- (800.7 °C) BP- (1,465 °C) Potassium Chloride (KCl): MP- (768 °C) BP- (1411 °C)}	Low melting and boiling point {Methane (CH_4): MP- (-182.5 °C), BP- (−162 °C) Nitrogen (N_2): MP- (−209.86 °C), BP- (−195.795 °C)}	High melting and boiling point {(Sodium (Na): MP- (97.794°C), BP- (882.940°C) Magnesium (Mg): MP- (650 °C), BP- (1,090 °C)}
Electrical conductivity	Will conduct electricity when molten or dissolved in water	Will not conduct electricity	Will conduct electricity when solid or molten
Magnetic property	Not magnetic	Not magnetic	Some are magnetic
Solubility issue in polar electrolyte (example: water)	Many are soluble in water	Are not soluble in water but are soluble in non polar solvents like ether	Are not soluble in water or any other solvent unless they undergo a reaction with them
Typical schematic example	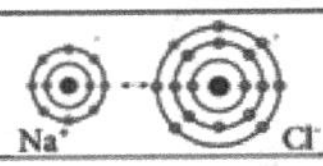		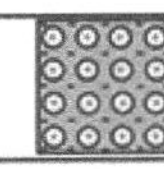
Typical examples	• Sodium Chloride (NaCl):Used to flavour food • Potassium iodide (KI) : to treat hyperthyroidism • Calcium oxide (CaO) : manufacturing of cement, paper, and high-grade steel.	• Water (H_2O) : drinking, cooking etc. • Oxygen (O_2): necessary for human life • Ethylene (C_2H_4): ripening fruit, making detergents etc.	• Copper (Cu): wiring electrical appliances • Diamond (C): cutting materials, as jewellery • Iron (Fe): To manufacture steel, for reinforcement in concrete etc.

1.7.5.1 Van der Waals BONDS

Van der Waals bonds are a result of an asymmetrical distribution of positive and negative charges inside each atom or molecule, which is created in a dipole. This type of force exists among all kinds of molecules and atoms. The strength of Van der Waals bonds is significantly weaker than ionic and covalent bonds. The Van der Waals forces of attraction exist between the inert gases and elementary noble gases. It is considered as a weakest chemical force and rapidly vanishes when intermolecular distance increases. The van der Waals forces are mainly affected by size and shape of atom/molecule. The strengths of Van der Waals forces commonly range from 0.4 $kJ.mol^{-1}$ to 4 $kJ.mol^{-1}$. There are three types of Van der Waals forces. This classification originates based on dipole–dipole, induced dipole–dipole, and induced dipole–induced dipole interactions.

The dipole–dipole interaction exists between molecules having permanent dipoles. The dipole–dipole interaction (*Keesom interaction*) results due to the electrostatic attraction between two permanent dipoles. Permanent dipoles are formed in asymmetric molecules where there are permanent positive and negative regions due to difference in electro-negativities of the constituent elements. The dipole–dipole interactions are weaker than ion-ion interactions and stronger than induced dipole–induced dipole interactions.

The induced dipole–dipole interactions result due to interaction between permanent dipoles and other molecules/atoms, which form an induced dipole. An induced dipole can be formed due to repulsive forces between a permanent dipole and electrons belonging to other molecules/atoms. The Van der Waals force originates from induced dipole–dipole interactions is also known as *Debye force.*

The induced dipole–induced dipole interactions result when the electrons in two adjacent atoms/molecules occupy positions that make the atoms/molecules form temporary dipoles. The Van der Waals force originates from induced dipole–induced dipole interactions is also known as *London dispersion force.* This type of force is believed to be the weakest of all Van der Waals forces.

1.7.5.2 Hydrogen bond

Hydrogen bonds occur when a hydrogen atom undergoes dipole-dipole attraction to an electronegative atom. Water is an excellent example of hydrogen bonding as shown in Fig. 1.18. Usually, hydrogen bonds occur between hydrogen and fluorine, oxygen, or nitrogen. Sometimes the bonding is intramolecular, or between atoms of a molecule, rather than between atoms of separate molecules (intermolecular). Hydrogen bonds are relatively stronger than Van der Waals forces but compared to primary bonds they are weak.

It is based on the fact that hydrogen being the smallest atom provides very little repulsion when interacting with highly electronegative atoms in other molecules and thus succeeds in forming partial bonds with them. This makes hydrogen bonds strong but weaker compared to primary bonds since the interactions here are permanent dipole interactions.

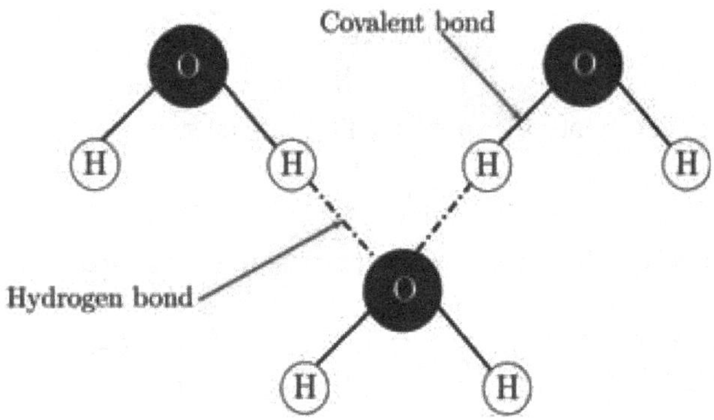

Fig. 1.18: Hydrogen bonding in water

The bond is between the hydrogen of one water molecule and the oxygen atoms of another water molecule, not between the two hydrogen atoms (a common misconception). How this works is that the polar nature of the water molecule, that is, each hydrogen atom experiences attraction to both the oxygen it's bound to and to the non-hydrogen side of the oxygen atoms of other water molecules. Hydrogen bonding in water results in the crystal structure of ice, making it less dense than water and able to float.

UNIT SUMMARY

1. Engineering materials are classified into metals, polymers and ceramics based on the chemical nature and atomic structure.
2. Unit cell is the smallest unit of a space lattice. We can create a space lattice by repeating unit cells.
3. Unit cells can be classified into two types, *primitive* and *non-primitive unit cells.* When motifs/basis are present only on the corner position of a unit cell, it is called a primitive *unit cell.*
4. In crystalline material, there are specific crystallographic directions and planes which contain a large number of atoms/ions. These specific directions and planes in a crystal is known as Miller indices.
5. Crystals are never perfect and always contain various types of defects. These defects are classified as point defect, line defects, surface defects and volume defects.
6. Bonding in solids may be classified as either primary or secondary bonding. The primary bonds are strong in nature and corresponds to ionic, covalent and metallic bonding.

EXERCISES

Multiple Choice Questions

1. Which of the following elements does not have BCC crystal structure?

(a) Fe (b) Mg

(c) Ni (d) W

2. Which of the following elements show covalent nature of bonding?

(a) Na (b) Si
(c) Fe (d) Cu

3. Which of the following elements does not have FCC crystal structure?
(a) Cu (b) Al
(c) Ti (d) Zn

4. Which of the following is in the correct order in case of inter atomic distance between two adjacent carbon atoms?
(a) C–C < C=C < C≡C (b) C=C > C–C > C≡C
(c) C≡C < C–C < C=C (d) C–C > C=C > C≡C

5. Which of the following statement(s) is/are true?
(a) The ionic radii for CN = 6 is 10% larger than CN = 4
(b) The ionic radii for CN = 6 is 3% smaller than CN = 8
(c) The distance between two neighbouring atoms increases with increase in temperature
(d) The ionic radii for CN = 6 is 3% larger than CN = 4

6. Which of the following is in the correct order in case of bond energies?
(a) C–C < C= C < C≡C (b) C=C > C–C > C≡C
(c) C≡C < C–C < C=C (d) C – C > C = C > C ≡ C

7. Choose the correct statement(s):
(a) The negative ion and its corresponding positive ion have same ionic radii
(b) The negative ion is larger than its corresponding atom
(c) The positive ion is smaller than its corresponding atom
(d) Fe^{3+} has lower ionic radii as compared to Fe^{2+}

8. Which of the following materials has a metallic bond?
(a) Na (b) Ar
(c) Ca (d) C (diamond)

9. What is the nature of bonding in ethane (C_2H_6)
(a) Ionic (b) metallic
(c) covalent (d) both metallic and ionic

10. Electrical and thermal conductivity of metals are commonly better than ionic and covalent bonded materials due to:
(a) the presence of positive and negative ions
(b) the presence of delocalised valence electrons
(c) lower potential energy value of metallic material
(d) All of these

11. Density of a material is a function of:
(a) atomic radii (b) temperature
(c) atomic weight (d) CN

12. If a particular material has higher interatomic forces of attraction, then which property will be higher?
(a) hardness (b) strength
(c) modulus of elasticity (d) elasticity

13. Which of the following is a group of metals having FCC crystal structure?

(a) Fe, Cu, Ti (b) Cu, Ni, Au
(c) Ni, Al, Ag (d) W, Ni, Mn

14. Which of the following represents a group of metals having BCC crystal structure?

(a) Si, Na, Li (b) Li, Cr, Fe
(c) Li, Cr, Ni (d) W, Na, V

15. Which of the following represents a group of metals having HCP crystal structure?

(a) Mg, Be, Zn (b) Mg, B, Ti
(c) Ti, Cd, Co (d) Zn, Fe, Ni

16. What is the crystal structure of pure iron between 1000°C and 1350°C?

(a) HCP (b) BCC
(c) FCC (d) BCC and FCC

17. Which of the following is a thermodynamically stable defect?

(a) Dislocation (b) vacancy
(c) crack (d) precipitate

18. Which crystal system has the lowest symmetry among the followings?

(a) Trigonal (b) cubic
(c) hexagonal (d) Triclinic

19. Which of the following is the close-packing arrangements Choose the correct statement(s):

(a) ----ABABABAB---- (b) ----ABCABCABC----
(c) ----ABCABABCABABC---- (d) ----AABCABABC-----

20. Which of the following defects is considered a one-dimensional defect?

(a) Dislocation (b) vacancy
(c) stacking fault (d) precipitate

ANSWERS															
1.	(b)	2.	(b)	3.	(c)	4.	(d)	5.	(a)	6.	(a)	7.	(b)	8.	(a)
9.	(c)	10.	(b)	11.	(b)	12.	(a)	13.	(b)	14.	(b)	15.	(a)	16.	(c)
17.	(b)	18.	(d)	19.	(a)	20.	(a)								

PROBLEMS

1. Distinguish between crystalline solids and amorphous solids. Write down at least two examples of both types of solids.
2. What is the difference between lattice and crystal?
3. Write a short note on the following: space lattice, motif or basis, unit cell, crystal structure, primitive cell, and non-primitive cell.
4. Define a unit cell of a space lattice. What lattice constants define a unit cell?
5. What are the 14 Bravais unit cells? Arrange them in the order of their symmetry.

6. What are the three most common metal crystal structures? List five metals (along with their lattice parameters) that have each of these crystal structures.
7. Draw the following lattices: simple cubic, BCC, FCC, and HCP. Determine the followings for each lattice:
   (a) Coordination number
   (b) Effective number of atoms
   (c) Packing factor and
   (d) packing efficiency
8. How many atoms are there in a unit cell of diamond? Determine the packing efficiency of diamond crystal.
9. Calculate the lattice parameter for the following BCC crystals at ambient temperature:
   (a) Molybdenum (atomic radius = 0.140 nm)
   (b) Iron (atomic radius = 0.124 nm)
   (c) Niobium (atomic radius = 0.143 nm)
10. Calculate the atomic radius of the following BCC crystals at ambient temperature:
   (a) Lithium (lattice parameter = 0.35092 nm)
   (b) Sodium (lattice parameter = 0.42906 nm)
   (c) Tungsten (lattice parameter = 0.316 nm)
   (d) Chromium (lattice parameter = 0.28905 nm)
11. Calculate the lattice parameter for the following FCC crystals at ambient temperature:
   (a) Aluminium (atomic radius = 0.143 nm)
   (b) Palladium (atomic radius = 0.137 nm)
   (c) Gold (atomic radius = 0.144 nm)
   (d) Platinum (atomic radius = 0.139 nm)
12. Calculate the atomic radius of the following FCC crystals at ambient temperature:
   (a) Silver (lattice parameter = 0.409 nm)
   (b) Gold (lattice parameter = 0.408 nm)
   (c) Copper (lattice parameter = 0.3615 nm)
13. Indicate the following points within a unit cell:
   (a) ¼, ¼, ¾ (b) 1/8, 1/8, 1/8
   (c) ¾, ¼, ¼ (d) 7/8, ¾, 1/3
14. List the atom positions for eight corners and one body centered atoms of the BCC unit cell.
15. List the atom positions for eight corners and six face-centered atoms of the FCC unit cell.
16. Indicate tetrahedral voids in BCC and FCC crystals. Also, write down their coordinates.
17. Indicate octahedral voids in BCC and FCC crystals. Also, write down their coordinates.
18. Determine the direction indices of the cubic direction between the position coordinates
   (a) (3/4, 0, 1/4) and (1/4, 1/2, 1/2).
   (b) (1, 1, 1) and (1, 1, 0)
   (c) (1/2, 0, 1) and (1/2, 0, 0)
   (d) (1/2, 1/2, 1/2) and (1/2, 0, 0)

19. Distinguish between lattice points and atomic sites in monoatomic FCC crystal and a NaCl crystal.
20. Calculate the distance in iron atom at room temperature between the center of a iron atom and the center of
    (a) Its nearest neighbour.
    (b) Its nearest tetrahedral void.
    (c) The nearest octahedral void.
21. Calculate the distance in NaCl between the center of a sodium ion and the center of
    (a) The nearest site that are identical.
    (b) Its nearest neighbour.
    (c) Its nearest positive ion.
    (d) Its second nearest negative ion.
22. Sketch the following direction vectors within a cubic unit cell:
    (a) [101]; (b) [102]; (c) [12$\underline{2}$]
    (d) [$\underline{2}$01] (e) [1$\underline{2}$0] (f) [1$\underline{1}$3]
23. Draw Miller indices of directions of a cubic system for the following family of direction:
    (a) <100>; (b) <110>; (c) <111>
24. Draw the crystal planes in a cubic unit cells that have following Miller indices:
    (a) (101); (b) (1$\underline{2}$0) (c) (221)
    (d) ($\underline{2}$01) (e) (1$\underline{1}$0) (f) (1$\underline{3}$1
25. A plane includes points at (0,0,0), (1,0,1) and (1,1/2,1/2). What are its Miller indices?
26. Find out the family of planes of the followings in a cubic system:
    (a) <100> (b) <110> (c) <111>
27. When a copper atom is located at the origin of an FCC unit cell, a small interstitial hole is centered at ¾, ¼, ¼. Identify three identical sites within an FCC unit cell for the hole positions. [Hint: The replication of points in FCC accompanies translation of $\pm\frac{a}{2},\pm\frac{a}{2},0$ of $\pm\frac{a}{2},0,\pm\frac{a}{2}$; and of $0,\pm\frac{a}{2},\pm\frac{a}{2}$ (in addition to the $\pm a, \pm a, \pm a$, translations).
28. How many atoms per mm2 are there at ?
    (a) On the (100) planes of Lead
    (b) On the (110) planes of Lead
    (c) On the (111) planes of Lead
29. How many atoms per mm2 are there at 200C
    (a) On the (100) planes of Iron
    (b) On the (110) planes of Iron
    (c) On the (111) planes of Iron
30. Which will have the largest interplanar spacing in platinum d200, d110, d111?
31. The attractive force between two opposite ions using Coulomb's law can be written as $F_{attractive} = -\frac{z_1 z_2 e^2}{4\pi\varepsilon_0 d^2}$ where Z_1 and Z_2 are number of electrons removed from or added to the atoms to form

ions, e is electron charge (1.6×10^{-19} C), ε_0 is permittivity of free space (8.85×10^{-12} $C^2/N.m^2$), and d is the interionic separation distance (m). Calculate the coulombic attractive force between a pair of Na^+ and Cl^- ions that just touch each other if the ionic radius of Na^+ and Cl^- ions are 0.095 nm and 0.181 nm respectively.

32. Using the above formula, determine the ionic radius of S^{-2} ion if the radius of Mg^{+2} ion and the attractive force between Mg^{+2} and S^{-2} ions are 0.065 nm and 1.49×10^{-8} N respectively.
33. X-rays of an unknown wavelength are diffracted by a gold sample (lattice constant ~ 4.0788 Å). The 2θ angle was 64.582° for the {220} planes. What is the wavelength of the X-ray used? [assume first order diffraction].
34. The interplanar distance between {110} planes for a BCC crystal system is 1.181 Å, which has been determined by using X-rays of wavelength 1.541Å. Comment on the order of Bragg's reflection observed in this case.
35. Why does n-pentane (C_5H_{12}) exist in liquid form whereas neopentane (C_5H_{12}) exist in gaseous form at room temperature?

PRACTICAL

Experiment title: Determination of crystal structure of a given material using X-ray diffraction.

Apparatus / Materials required: Metal sample, and X-ray diffractometer.

Objectives: (a) Determining the lattice parameter.

(b) Identifying the crystal structure using the given XRD pattern.

Theory:

For the principles and fundamentals of X-ray diffraction technique, the students can refer to the book "Elements of X-ray diffraction" by B.D. Cullity. To analyze the diffraction pattern, the conditions for the constructive and destructive interference for a cubic crystal system is shown in Table below.

Bravais lattice	Constructive interference	Destructive interference
BCC	(h+k+l) = Even	(h+k+l) = odd
FCC	(h,k,l) all odd or all even	(h,k,l) not all odd or all even

Observation:

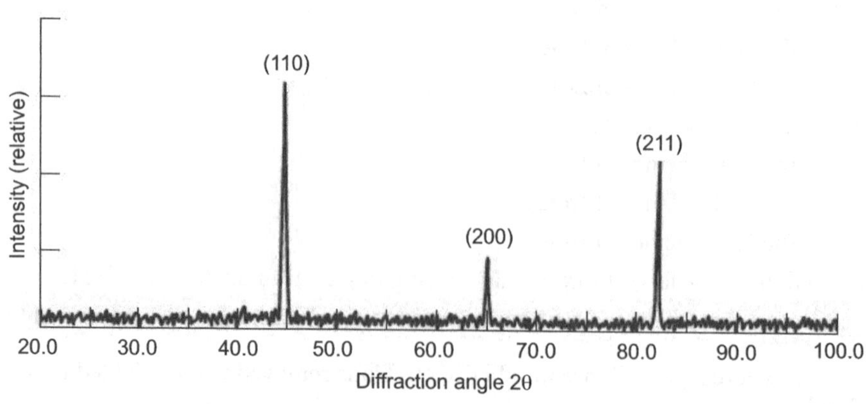

Diffraction pattern of an unknown metal

Result:

Analysis of the diffraction pattern lines.

Peak.	2θ	sinθ	Interplanar spacing	h k l	Lattice parameter
1.					
2.					
3.					

Conclusion:

The lattice parameter for the given metal is .... nm

The crystal structure of the given metal is ....

KNOW MORE

Students can refer to advanced concepts on symmetry, various symmetry operations, space groups and point groups.

REFERENCES AND SUGGESTED READINGS

1. William D. Callister, Jr., David G. Rethwisch, Materials Science and Engineering: An Introduction, John Wiley & Sons, 9th edition, 2014.
2. V. Raghavan, Physical Metallurgy Principles and Practice, PHI Learning Private Limited, Second Edition 2009.
3. R.E. Smallman, A.H.W. Ngan, Modern Physical Metallurgy, Butterworth-Heinemann Ltd., 8th edition, 2014.
4. R. Abbaschian, L. Abbaschian, R.E. Reed-hill, Physical Metallurgy Principles, Cengage Learning India Pvt. Ltd., 4th edition, 2010.

Dynamic QR code for further reading

2 Phase Diagrams, Ferrous metals, and their Alloys

UNIT SPECIFICS

Through this unit the following aspects have been discussed:

- Isomorphs, eutectic and eutectoid systems; Iron-Carbon binary diagram; Iron and Carbon Steels; Flow sheet for production of iron and steel; Iron ores–Pig iron: classification; composition and effects of impurities on iron;
- Cast Iron: classification, composition, properties and uses; Wrought Iron: properties, uses/applications of wrought Iron; comparison with cast iron;
- Wrought iron, mild steel and high carbon steel; standard commercial grades of steel as per BIS and AISI;
- Alloy Steels – purpose of alloying; effects of alloying elements – Important alloy steels: Silicon steel, High Speed Steel (HSS), heat resisting steel, spring steel;
- Stainless Steel (SS): types of SS, applications of SS – magnet steel – composition, properties and uses.

LEARNING OBJECTIVES

The practical applications of the topics are discussed for generating further curiosity and creativity as well as improving problem solving capacity in Material Science and Engineering.

Besides giving many multiple-choice questions as well as questions of short and long answer types marked in two categories following lower and higher order of Bloom's taxonomy, assignments through a few numerical problems, a list of references and suggested readings are given in the unit so that one can go through them for further advanced knowledge.

This section also highlights the examples of some interesting facts, analogy, history of the development of the subject focusing on the salient observations and finding, timelines starting from the development of the concerned topics up to the recent time, applications of the subject matter for industrial applications on variety of aspects and/or our day-to-day real life and finally inquisitiveness associated with the topics of the unit.

RATIONALE

This unit helps students to get a primary idea about the relationships between phases in equilibrium in a system as a function of temperature, composition, and pressure. It explains the Phase rule, unary phase diagram, binary phase diagram and its industrial application. All these basic aspects are relevant to understand the equilibrium phases present in a specific metallic system at a given temperature and pressure. It then clearly explains development of iron-based alloys (Ferrous metals and its alloys). Important ferrous alloys along with their properties and application have been discussed to further enhance the knowledge of the students. The contribution of Indian Metallurgy to develop advanced ferrous materials is also discussed.

Its practical applications are related to the determining the equilibrium phases, process parameter optimization during metal working, and materials design and selection. Finally, student will be capable to correlate the application of phase diagram in respective industries, and/or to acquire the knowledge of phase transformation of common ferrous system.

PRE-REQUISITES

Physics, Chemistry, and Thermodynamics (Class XI, XII)

UNIT OUTCOMES

List of outcomes of this unit is as follows:

U2-O1: To describe basics of phase diagram, and phase rule.

U2-O2: To describe the Fe-C phase diagram.

U2-O3: To describe the production of pig iron and various iron-based alloys.

U2-O4: To understand the role and application of phase diagram in materials selection and design.

Unit-2 Outcomes	EXPECTED MAPPING WITH COURSE OUTCOMES (1- Weak Correlation; 2- Medium correlation; 3- Strong Correlation)				
	CO-1	CO-2	CO-3	CO-4	CO-5
U2-O1	3	3	3	2	2
U2-O2	1	3	2	2	1
U2-O3	2	3	3	1	2
U2-O4	3	3	3	1	2

2.1 INTRODUCTION AND BASIC TERMINOLOGY

This unit helps students to get a primary idea about the relationships between phases in equilibrium in a system as a function of temperature, composition, and pressure. It explains the Phase rule, unary phase diagram, binary phase diagrams and its industrial application. All these basic aspects are relevant to understand the equilibrium phases present in a specific metallic system at a given temperature and pressure. Further, a brief introduction to steel making, the various processes, compositions, heat treatments and the challenges faced by this mega industry within the realms of the "science and technology" of steelmaking have been discussed which is extremely vast and may not be easily understood within one chapter. This chapter introduces to the reader the intricacies of this age-old practice and research continues up to the modern era of technology development.

A phase diagram is basically the graphical representation of different phases present in the materials system at various pressures, temperatures, and compositions. These diagrams are drawn by utilizing the different equilibrium conditions and are widely used by scientists and engineers/industrialists to predict and understand the different aspects and behavior of the material. Phase diagrams are very useful in predicting phase transformations that occur during materials processing under different process conditions (composition, temperature, pressure etc.) and their resulting microstructure in equilibrium. The phase diagram shows the stable states or equilibrium phases of an alloy, i.e. those which have the least free energy under given thermodynamic conditions. To utilize the phase diagram and its application in materials science and engineering, following technical terms need to be understood:

Grain: A grain is a single crystal within which the atomic lattice and its orientation are continuous. i.e., in any one grain, all atoms are arranged in one particular orientation and one particular pattern.

Grain boundary: A grain boundary is an interface between two grains. A polycrystal is generally made up of several grains separated by grain boundaries. The orientations of these grains change abruptly across the grain boundaries.

Microstructure: In most materials, the sizes of the grains are of the order of microns ($\sim 10^{-6}$ m) which are examined with the help of an optical microscope. However, study of nanosized grains ($\sim 10^{-9}$ m) are also possible using electron microscopy. The arrangement of these grains and the therefore the structure of the surface of the material revealed by an optical / electron microscope is called Microstructure as shown in Fig 2.1.

Microstructure consists of grains, grain boundaries, and other common types of defects present in materials.

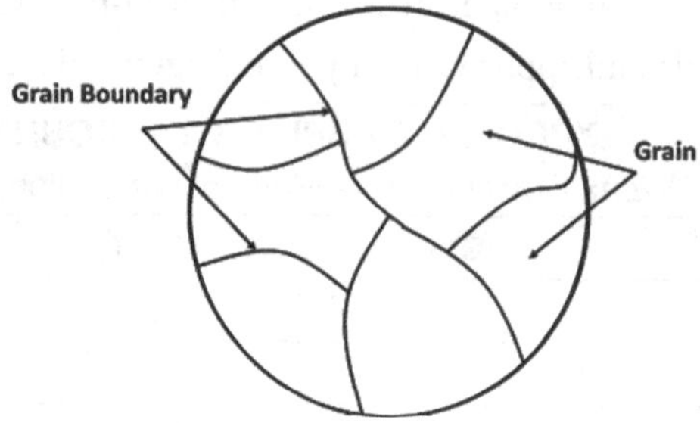

Fig 2.1: Schematic representation of a microstructure consisting of grains and grain boundaries.

Component: The component of a system can be defined as the independent species that comprises the system. The component or constituent of a system may be elements, ions, or compounds. Consequently, a pure metal represents a single component system, an alloy of two metals, a two-component system, and so on. These components will not dissociate into other species during the entire range of metal working processes. Chemical compound such as cementite (Fe_3C) can be regarded as component as it does not dissociate in the entire temperature range considered for metallurgical processing.

Solid Solution: A homogeneous mixture of substances in the solid form is called a solid solution. The solution consists of molecules or atoms which share crystal lattice with each other, as in different metal alloys. For e.g., steel is a solid solution of carbon and iron which is used for construction purposes. There are two types of solid solutions namely (a) substitutional solid solution; and (b) interstitial solid solution.

Substitutional solid solution: The substitutional solution is a solution in which the minor components atoms (solute atoms) substitute or replace the major atoms (solvent atoms) in their lattice sites.

Interstitial solid solution: In this solution, the solute atoms go into the interstitial sites i.e., spaces between the solvent atoms.

Phase: A phase is defined as a unique portion of a system which is different in composition and/or microstructure from the other region. The region could be of a single component or a mixture of two or more components. Each phase will have a distinct property due to different chemical composition and its associated crystalline behaviours. Therefore, a phase is a physically distinct, chemically homogenous and mechanically separable region of a system. At the end of the material development process, the entire developed product would consist of the final phase (single phase or multiphase system) that was intended to be fabricated.

Phase transformation: A phase transformation is the physical process in which the phases present in a system undergo changes or there is change in the state of the material as a function of temperature and/or pressure. The desired structure and thus the final property of a material is achieved by understanding the phase transformation mechanism and associated kinetics.

Equilibrium: A system is said to be under equilibrium when its free energy is low under some defined conditions of pressure, temperature, and composition. A system is said to be in equilibrium when its thermodynamic properties do not change with time. An illustration of different stages is shown in figure 2.2.

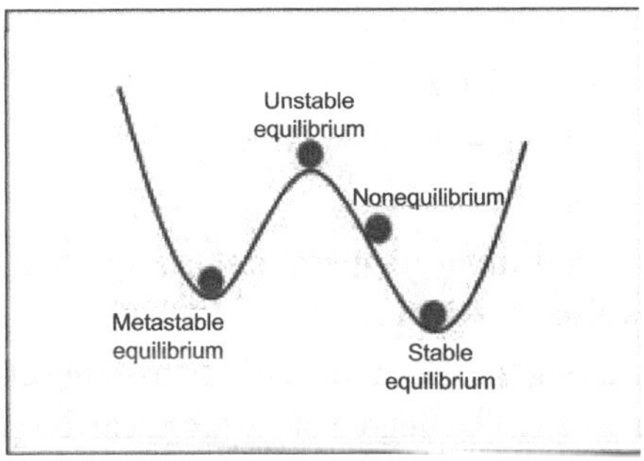

Fig. 2.2: Schematic illustrating the different states of equilibrium

In Fig. 2.2, the free energy has a local minimum at the metastable equilibrium 'state A'. Due to the small fluctuations, the system again gains the same state and often tries to reach a much stable state. 'State B' corresponds to the stable equilibrium state, where the free energy is at a global minimum. When the system moves from state A to state B, the free energy rises and reaches to a local maximum. The system becomes unstable due to fluctuations at that unstable equilibrium state; hence, it is short-lived. During the movement between the equilibrium states, the system undergoes a series of unstable or non-equilibrium states where the properties of the entire system get frequently changed. A system is said to be in thermodynamic equilibrium condition when mechanical equilibrium, chemical equilibrium and thermal equilibrium coexists in that system at a particular time.

2.2 PHASE RULE

J. W. Gibbs gave an equation which determines the number of phases present in a given system at particular set of thermodynamic conditions. The equation is commonly called as Gibbs phase rule. The phase rule equation establishes the quantitative relationship between the number of phases P, number of degrees of freedom and the number of components C in a system:

$$P + F = C + 2$$

The components C present in the system could be a compound, an alloying element, or a solution in the system. The degree of freedom, F, is the number of variables (composition, temperature, and pressure) that can be altered without varying the number of phases in the system and 2 is written to represent the number of variables (i.e., pressure and temperature).

If the degree of freedom is zero, then the internal and external factors (concentration, temperature, pressure) of the system cannot be changed without changing the number of phases. Similarly, if the degree of freedom is unity, then the internal and external factors of the system can be changed within certain limit without causing changes in the number of phases.

We know the fact that water (H_2O) can sustain in all three phases i.e., solid, liquid, and vapor. The different phases at different conditions are shown with the help of pressure-temperature diagram in Fig. 2.3. We can see from the figure that there exists a junction called as triple point (O). At this point, all the three phases of water coexist at the same time. Solid (ice) and liquid (water) exists along the freezing line (shown as green line), whereas vapor (steam) and liquid (water) exist along the vaporization line (shown as blue line). It is interesting to note that solid (ice) and vapour (steam) can also coexist (shown as red line) at lower pressure and sub-zero temperature.

Since at triple point, the three phases coexist together , thus there is only one component i.e., C = 1. The number of degrees of freedom can be calculated as:

$$P + F = C + 2$$
$$3 + F = 1 + 2$$
$$F = 0$$

Since none of the variables (pressure or temperature) can be changed and the three phases are still in the balance, the triple point is also called invariant.

Now, let us consider a point along the green line i.e., the freezing curve. On this curve, the solid and liquid phases can coexist together i.e., . The degree of freedom can be calculated as:

$$P + F = C + 2$$
$$2 + F = 1 + 2$$
$$F = 1$$

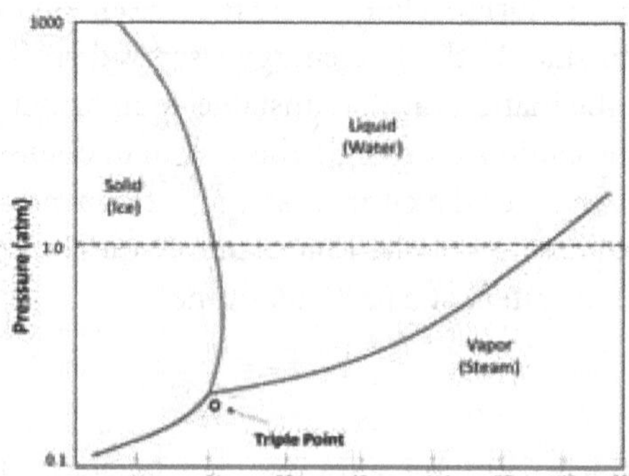

Fig 2.3: Pressure-temperature diagram of H_2O.

Here, the one value of F indicates that by changing either pressure or temperature, the system will have two phases. If the temperature changes, the pressure has to be adjusted to maintain the two phases.

Finally, let us consider a point within the liquid region. Since, there is only one phase, thus . The degree of freedom can be calculated as:

$$P + F = C + 2$$

$$1 + F = 1 + 2$$

$$F = 2$$

The results display that two variables can be independently changed, and the system will remain in one phase.

In most metallurgical systems, the pressure is generally kept constant which reduces the number of variables by one. Thus, the phase rule becomes:

$$P + F = C + 1$$

We can further classify phase diagram based on number of components present in the system.

2.3 UNARY PHASE DIAGRAM

The unary phase (single component) diagram consists of a one-component system in which the component is held constant (i.e., the phase diagram is for a pure substance); this means that pressure and temperature are the variables. The unary phase diagram is also called a pressure-temperature (or P – T) diagram which is represented as a 2D plot of pressure versus temperature. This type of diagrams can be seen in Fig. 2.4 (for metal) and Fig. 2.5 (for nonmetal). Figures 2.4a and 2.4b show the effect of temperature and pressure on the phase transformations of pure iron. Iron changes its crystal structure from BCC to HCP at room temperature on application of very high pressure (see Fig. 2.4b). Fig. 2.5 shows the effect of pressure on the equilibrium temperatures for carbon. There are two triple points visible in case of pure carbon.

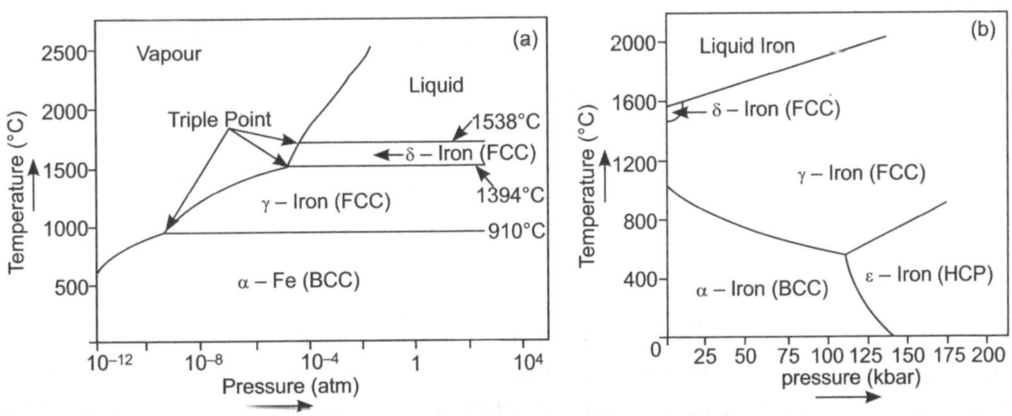

Fig. 2.4: Pressure-temperature diagram for pure Iron at (a) lower pressure range, and (b) higher pressure range (1atm ~ 1.013 bar pressure).

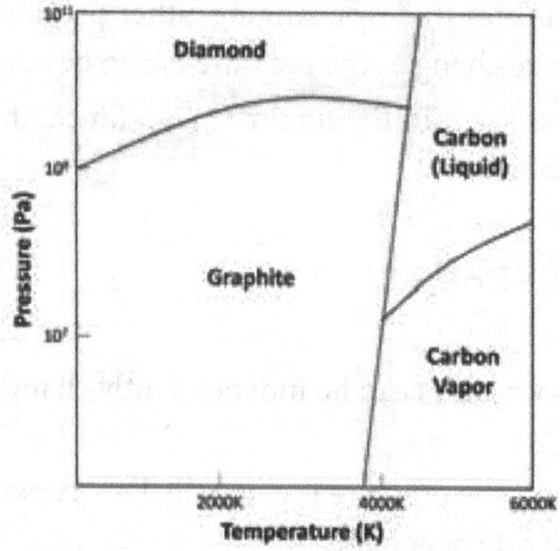

Fig. 2.5: Pressure-temperature diagram for pure carbon.

2.3.1 Cooling Curves

Cooling curves are used to determine the phase transitions of metals and alloys. A cooling curve is generally obtained by recording the temperature versus time plot. Let us consider the example of the cooling curve for pure metal as shown in Fig. 2.6. The curve AB represents the cooling curves after the molten metals (in liquid state) are poured into a container. There is a need to provide undercooling to start the liquid to solid transformation. The amount of undercooling required depends on several factors. Point B represents the freezing point where solidification and the curves become flat, and it remains flat up to point C until the solidification becomes complete.

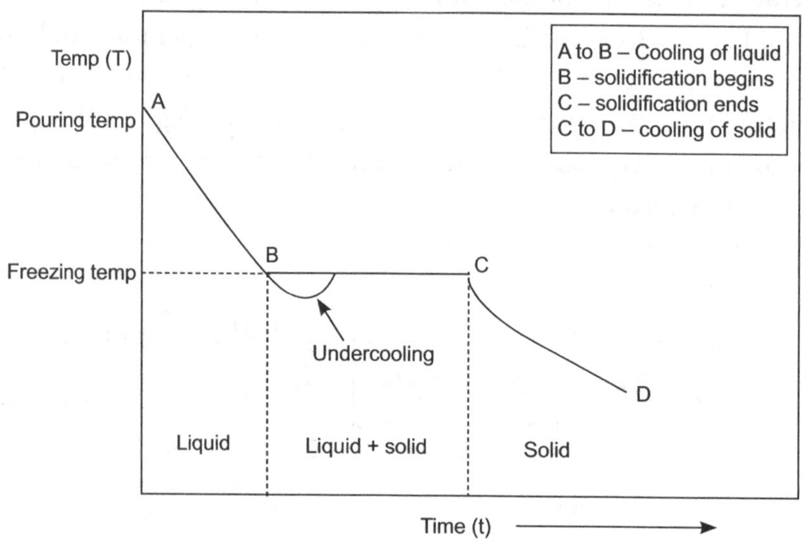

Fig. 2.6: Cooling curve of a pure metal

The metal is in the form of a mixture of solid and liquid phases between point B and C. After point C, the cooling curve again shows a drop in temperature with time as represented by segment CD in the figure.

2.4 BINARY ISOMORPHOUS SYSTEM

A binary isomorphous system is a system which comprises of a mixture of two metals. It basically consists of two components as each element is considered as a separate component in an alloy. We can omit one variable and the phase rule can be written as:

$$P + F = C + 1$$

Let us understand the construction of these diagrams. In general, the solidification of metal occurs at constant temperature whereas it occurs over a wide range of temperature for an alloy. Consider an example of a binary isomorphous system i.e. Copper-Nickel alloy system as shown in Fig. 2.7. As we increase the content of Ni in the alloy, the freezing temperature increases and begins from A (where only pure Cu is present), A1, A2, A3, up to A4 (where only pure Ni is present) and ends at temperatures B (where only pure Cu is present), B1, B2, B3, up to B4 (where only pure Ni is present). It is always recommended to take more data points to get a more accurate and reliable phase diagram. The line obtained after joining the points from A to A4 is called as liquidus line and the line obtained after joining the points B to B4 is called as the solidus line. Thus, the locus of freezing start temperatures and freezing finish temperatures are known as liquidus line and solidus line respectively. Above the liquidus line, the alloy will be in liquid form whereas below the solidus line, the alloy remains in solid state. The intermediate zone between both the lines is also referred to as the Mushy zone where both the phases exist in equilibrium.

Thus, with the help of a cooling curve i.e. temperature-time diagram, we can easily construct a phase diagram after understanding and identifying the freezing zones i.e., the starting and termination points. However, even after the solidification, there is a change in the structure of the alloy system. Then, other methods such as x-ray, dilatometry methods, electrical conductivity, etc have to be explored. The above methods make it a tedious task to construct a phase diagram of such materials, due to the variable working conditions.

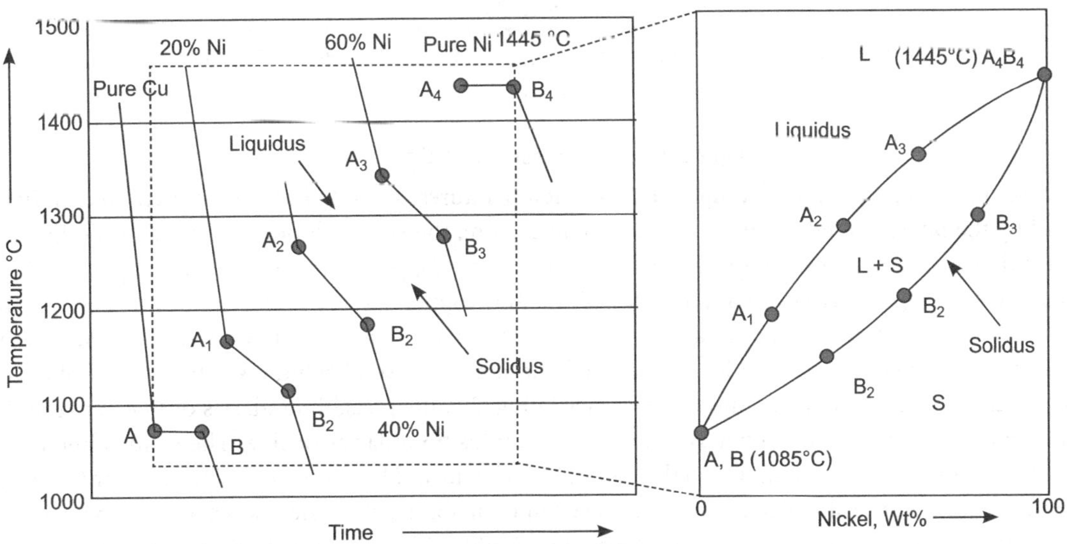

Fig 2.7: Phase diagram construction from the cooling curves

When the system consists of components having a similar crystalline structure, the components form a series of solid solutions due to the complete solubility of the components in their solid form. The copper-nickel phase diagram is a common isomorphous system which is shown in Fig. 2.8. The y-axis corresponds to the temperature whereas the x-axis represents the composition. The extreme right end of the composition scale represents 100 wt% Ni (0 wt% Cu) whereas the extreme left end represents 0 wt% Ni (100 wt% Cu). It can be seen that three different regions of different phases are present in the phase diagram i.e., solid-solution region (), liquid region (), and a mixture of solid and liquid () region. The melting points of pure Cu and Ni are unique points in Cu-Ni phase diagram as the degree of freedom is zero at both points. Student is suggested to apply phase rule in different region of Fig. 2.8 to determine the degree of freedom.

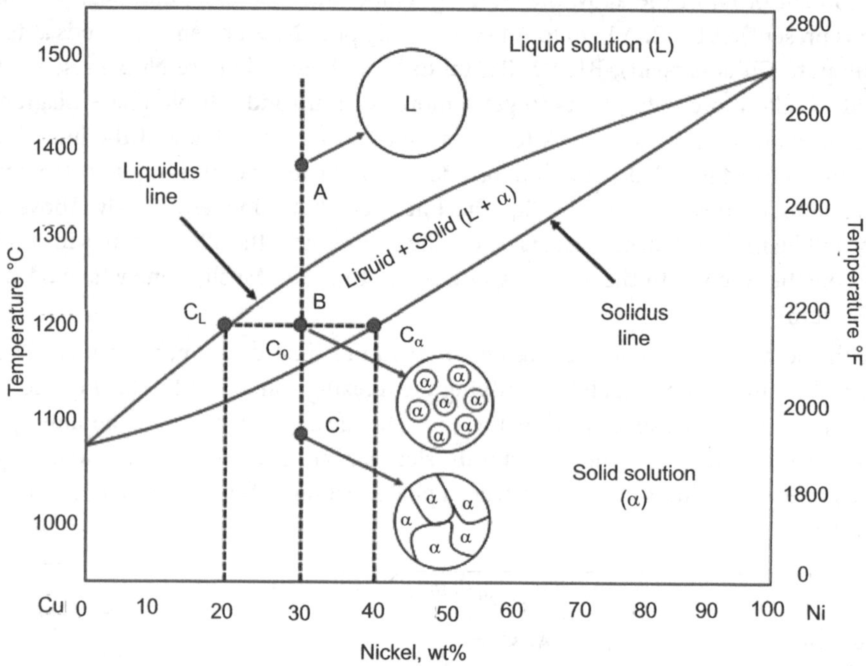

Fig 2.8: Copper-nickel phase diagram

Over the different range of compositions and temperatures, several phases are present in each field, bound by the phase-boundary lines. At lower temperature, the solid solution (α) exists in the form of substitutional solid solution comprising of both nickel and copper atoms whereas the liquid field (L) exists at high temperature consisting of both nickel and copper. At any given composition, when the alloy freezes, the nickel and copper atoms are mutually mixable and soluble with each other, thereby exhibiting a complete solid solubility. We generally designate the solid solutions with the help of Greek letters such as α, δ, ε, β, γ etc. As discussed earlier in Fig. 2.7, the phase-boundaries or line between the liquid and solid regions are known as liquidus and solidus boundaries or line. The solidus line is the curve separating the solid-solution and the two-phase of solid and liquid region. The line represents the maximum temperature at which all the atoms exist in their solid state. The line separating the liquid () and mixture of liquid plus solid (L + α) is termed as liquidus line. This line indicates the minimum or lower temperature at which any component exists in its molten state.

In a simple statement, the solidus line represents the series of points where the alloy begins melting on heating or finishes freezing on cooling whereas the liquidus line represents the series of points where the alloys finishes melting on heating and begins to freeze on cooling.

To attain complete solubility, *Hume-Rothery rules* needs to be fulfilled which is stated as:

1. The size difference of the atoms should not be greater than15% among the given two elements.
2. No electronegativities difference should be there in the given elements i.e., they should not form any compound with each other.
3. Each element of the solution must have similar kind of crystal structure.
4. The valency of each element must be same.

In this case, both copper and nickel have the FCC crystal structure, have nearly identical atomic radii and electronegativities, and have similar valences. The term isomorphous implies complete solubility in both the liquid and solid states. Most alloys do not have such simple phase systems. Typically, alloying elements have significant differences in their atomic sizes and crystalline structures, and so, the mismatch forces the formation of a new crystal phase that can more easily accommodate alloying elements in the solid state.

The liquidus and solidus lines intersect at the two composition extremities, that is, at the temperatures corresponding to the melting points of pure copper (1084°C) and pure nickel (1455°C). Since pure metals melt at a constant temperature, pure copper remains a solid until its melting point of 1084oC is reached on heating. The solid-to-liquid transformation then occurs, and no further heating is possible until the transformation is complete. However, for any composition other than the pure components, melting will occur over a range of temperatures between the solidus and liquidus lines. For example, on heating a composition of 50wt% Cu-50 wt% Ni (dotted lines indicated in figure to show the composition), melting begins at approximately (1250°C (2280°F), and the amount of liquid increases until about 1315oC (2400°F) is reached, at which point the alloy is completely liquid. A binary phase diagram can be used to determine three important types of information: (1) the phases that are present, (2) the composition of the phases, and (3) the percentages or fractions of the phases.

During the rapid solidification of many alloys, the compositional gradient is created, and this leads to formation of cored structure. Cored structure is a casting defect which can be eliminated by homogenizing the cast alloy for long times at a higher temperature. The selected temperature should be less than the melting point of lowest melting phase in the alloy.

Similarly, Fig. 2.9 represents the binary phase diagram of Al_2O_3-Cr_2O_3. The left end of the composition axis represents the pure Al_2O_3 as a component whereas the right end represents pure Cr_2O_3 as another component. The composition is read as 0% Cr_2O_3 from the left and 100% Cr_2O_3 to the right. The single-phase region i.e., liquid and solid are present in the diagram, separated by a two-phase region consisting of solid and liquid.

2.4.1 Prediction of Phases

The phases that are present can be determined by locating the temperature-composition point on the diagram and noting the phase(s) present in the corresponding phase field. For example, an alloy of composition 30wt%Ni-70wt%Cu at 1315°C would be located at point A in Fig. 2.8. Since this point lies totally within the liquid field, the alloy would be

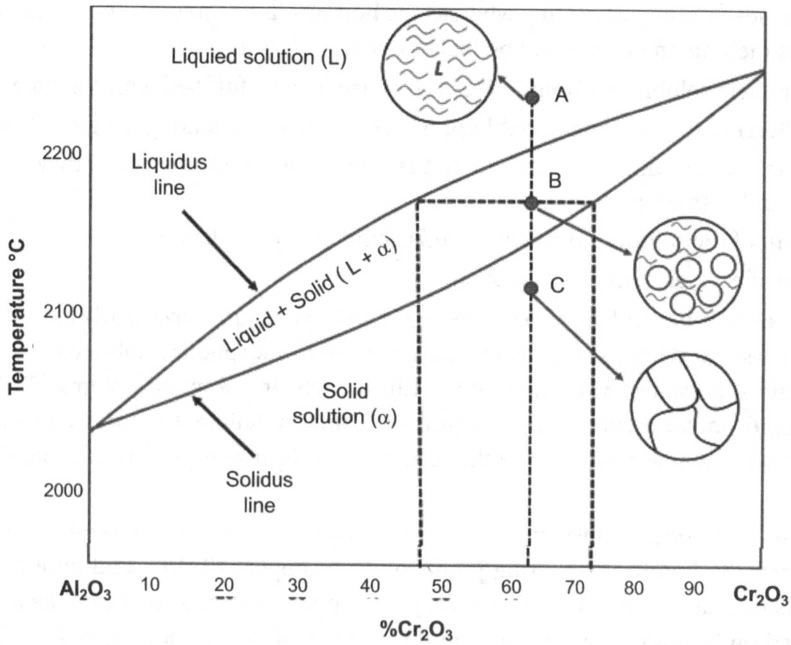

Fig. 2.9: $Al_2O_3 - Cr_2O_3$ phase diagram

a liquid. The same alloy at 1095°C, designated point C, is within the solid solution, , field; only the single a phase would be present. On the other hand, a 30wt.% Ni-70wt.% Cu alloy at 1190°C (point B) would consist of a two-phase mixture of solid solution, , and liquid, .

2.4.2 How to Determine Chemical Compositions of Phases?

To determine the composition of the phases present, locate the point on the phase diagram. If only one phase is present, the composition of the phase is the overall composition of the alloy. For example, for an alloy of 30wt.%Ni–70wt.%Cu alloy at 1095°C (point C in Fig. 2.8), only the phase is present, and the composition is 30wt%Ni-70wt%Cu. For an alloy with composition and temperature coordinates located in a two-phase region, the compositions of the phases can be determined by drawing a horizontal line, referred to as a tie line, between the two-phase boundaries at the temperature of interest. Then, drop perpendicular lines from the intersections of each boundary down to the composition axis and read the compositions. For example, again consider the 30wt.%Ni-70wt.%Cu alloy at 1190oC, located at point B in Fig. 2.8 and lying within the two-phase, α + L, field. The perpendicular line from the liquidus boundary to the composition axis is 20wt%Ni-80wt%Cu which is the composition, C_L, of the liquid phase. In a similar manner, the composition of the solid-solution phase, , is read from the perpendicular line from the solidus line down to the composition axis, in this case 40wt%Ni-60wt%Cu.

2.4.3 Prediction of Amounts of Phases (The Lever Rule)

The percentages or fractions of the phases present at equilibrium can also be determined with phase diagrams. In a single-phase region, since only one phase is present, the alloy is comprised entirely of that phase; that is, the phase fraction is 1.0 and the percentage is 100%. From the previous example for the 30wt%Ni-70wt%Cu alloy at 1095°C (2000°F) (point C in Fig. 2.8), only the phase is present, and the alloy is 100% . If the composition and temperature position is located within a two-phase field, a horizontal tie line must be used in conjunction with the lever rule. The lever rule is a mathematical expression (Eqn. 1) based on the principle of conservation of matter. First, a tie line is drawn across the two-phase region at the composition and temperature of the alloy as shown in Fig 2.10. The fraction of one phase is determined by taking the length of the tie line from the overall alloy composition to the phase boundary for the other phase and dividing by the total tie line length. The fraction of the other phase is then determined in the same manner. If phase percentages are desired, each phase fraction is multiplied by 100. When the composition axis is scaled in weight percent, the phase fractions computed using the lever rule are mass fractions – the mass (or weight) of a specific phase divided by the total alloy mass (or weight). The mass of each phase is computed from the product of each phase fraction and the total alloy mass.

Again, consider the 60wt%A–40wt%B alloy system at a particular temperature which is shown as horizontal dotted line in Fig 2.10. This alloy contains both the solid, α (45 wt.%A–55 wt.%B), and the liquid, L (70 wt.%A–30 wt.%B) phases at this temperature. The same tie line (dotted line) that was used for determination of the phase compositions can again be used for the lever rule calculation. The overall alloy composition located along the tie line is C_α–C_L, or 55-30 wt%. The weight percentage of liquid present at that temperature is then:

$$\text{wt\% of liquid} = \frac{C_a - C_{L0}}{C_a - C_L} \times 100 = \frac{55-40}{55-30} \times 100 = 60\% \quad (1)$$

Likewise, the amount of solid present at that temperature is:

$$\text{wt\% of solid} = \frac{C_0 - C_L}{C_a - C_L} \times 100 = \frac{40-30}{55-30} \times 100 = 40\% \quad (2)$$

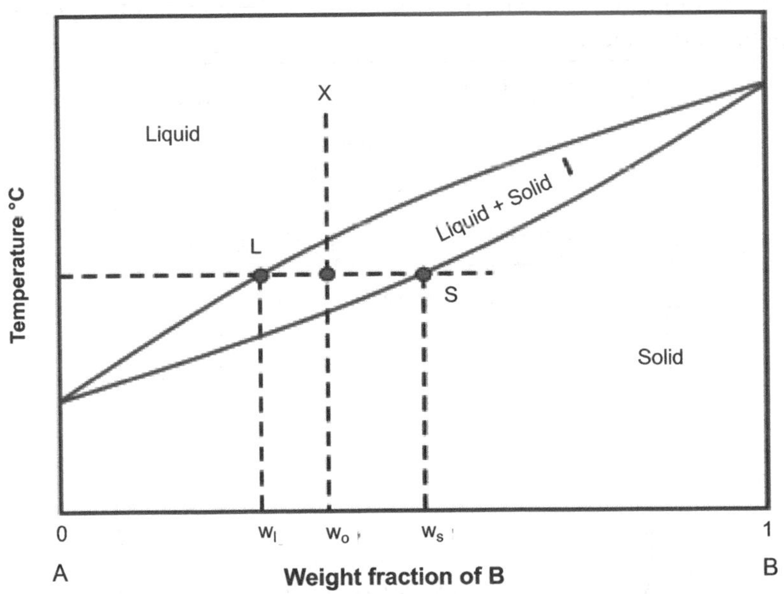

Fig. 2.10: A hypothetical phase diagram of two metals A and B completely soluble in each other.

2.1. A Cu-Ni alloy contains 47% Cu and 53% Ni and is at 1300°C. With the help of Fig. 2.8, Find:

(1) The wt% of Cu in the liquid and solid phases at this temp.

(2) How much is the wt% of the alloy is liquid and how much is solid at 1300oC?

Answer 2.1:

(1) At 1300°C, the tie line intersection gives 38% Cu in solid phase and 62% Cu in liquid phase.

(2) Using the lever rule at the tie line,

w_0 = 53% Ni, w_l = 45% Ni, and w_s = 58% Ni

$$\text{Weight fraction of Liquid: } = X_l = \frac{w_s - w_0}{w_s - w_l} = \frac{58-53}{58-45} = \frac{5}{13} = 0.38$$

wt% of liquid = 38%

$$\text{Weight fraction of Solid: } = X_s = \frac{w_0 - w_l}{w_s - w_l} = \frac{53-45}{58-45} = \frac{8}{13} = 0.62$$

wt% of solid = 62%.

2.5 EUTECTIC REACTION

Most alloy systems do not have such simple phase diagrams as the previous completely isomorphous system. One common type of binary phase diagram is a eutectic system, where the term eutectic is taken from the Greek for "easy or well melting." A eutectic reaction is a reaction where a liquid phase will form into two solid phases at a particular temperature. A common eutectic phase diagram is shown in Fig. 2.11. Eutectic systems form when alloying additions cause a lowering of the liquidus lines from both melting points of the pure elements.

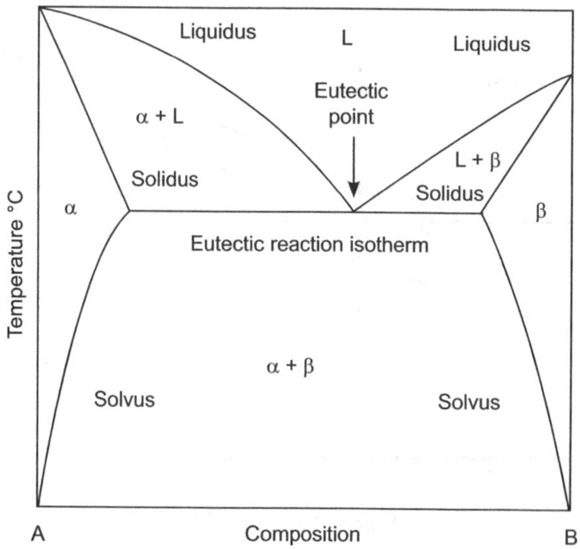

Fig 2.11: Phase diagram displaying eutectic reaction

At a specific composition, there is a minimum melting point, where the mixed solid-liquid phase regions (L + α and L + β) vanish. This is the eutectic point, which defines an alloy composition that has the lowest melting point of the A–B system. The eutectic composition also solidifies completely at a single temperature that is referred to as an invariant point. In a eutectic reaction, the liquid freezes to form two solid solutions:

$$\text{Liquid (L)} \xrightarrow{\text{Freeze}} \text{Solid } (\alpha) + \text{Solid } (\beta)$$

An alloy having a composition to the left of the eutectic point is called a hypoeutectic alloy (hypo is from the Greek meaning "less than"), while an alloy to the right of the eutectic point is a hypereutectic alloy. The phase formed above the eutectic temperature is known as proeutectic phase.

2.6 PERITECTIC REACTION

A peritectic reaction is a reaction where a solid phase and liquid phase will together form a second solid phase at a particular temperature and composition.

The difference between eutectic and peritectic is that eutectic describes the thermodynamic equilibrium conditions where a liquid coexists with two solid phases while peritectic describes the isothermal reversible reaction of a liquid phase and a solid phase to form a second solid phase during cooling. Schematic of peritectic reaction for carbon steel is shown in Figure 2.12. In a peritectic reaction, the liquid freezes and together with solid will form different solid solution:

$$\text{Liquid (L)} + \text{Solid } (\delta\text{-Fe}) \xrightarrow{\text{Freeze}} \text{Solid } (\gamma - \text{Fe}).$$

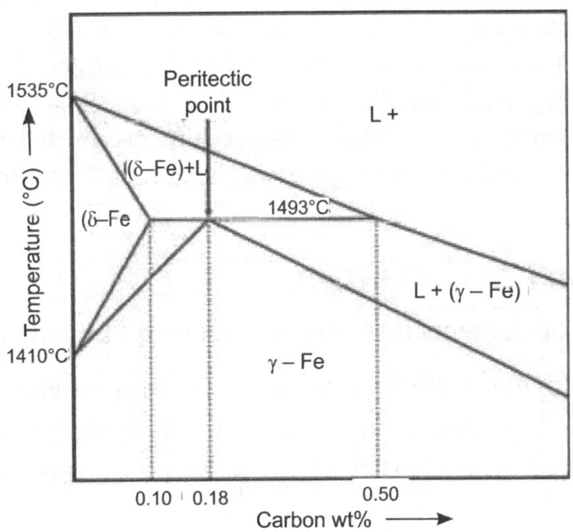

Fig. 2.12: Schematic of peritectic reaction for carbon steel

In peritectic reaction surrounding or enveloping occurs when the product phase (γ-Fe) surrounds the unreacted (δ - Fe) phase as parent phases (L and δ - Fe) get separated and are no longer in contact with each other.

2.7 FERROUS METALS

2.7.1 Types of Iron

Iron serves as the base for some of the most significant engineering alloys. Iron and its products manufacturing is a highly technical matter which involves good judgement of the combination of compositions, production methodologies and the desired properties for the intended applications. A relatively small amount of commercially pure iron is used in comparison with that used in alloys. *Ingot iron, electrolytic iron, hydrogen purified iron* and *carbonyl iron* are considered as commercially pure iron. They are relatively soft. The strength of iron can be raised by strain hardening (i.e. cold working) or by alloying with carbon and other elements. The trade name of Armco Ingot Iron is extensively used in roofing, ducts, drainage culverts, and as a base for porcelain enamel and is used in refrigerator cabinets, sign panels, stoves, etc.

There are three allotropic forms of iron, namely delta (δ, gamma (γ), and alpha (α). The pure iron is ferromagnetic at room temperature. Based on magnetic characteristics, there are two forms of –iron and the curie temperature of pure iron is 768°C. Iron is ferromagnetic below 768°C and is paramagnetic in the temperature range of 768oC to 910°C.

Prior to the extensive use of modern steel making process, wrought iron was the main metal for construction purposes. The pudding process was used to manufacture wrought iron which is mechanical mixture of very pure iron and a silicate slag (2 to 4%). Wrought iron is never cast. In fact, shaping of wrought iron is accomplished by hammering, forging, or pressing.

Now a days, wrought iron is known as an iron alloy with very low carbon content with respect to cast iron/plain-carbon steel. It is soft, ductile, magnetic, and has high elasticity and tensile strength. Although wrought iron exhibits properties that are not found in other forms of ferrous metal, it lacks the carbon content necessary for hardening through heat treatment. Wrought iron may be welded in the same manner as mild steel, but the presence of oxides or inclusions will provide defective results. Wrought iron contains less than 0.10% of carbon, less than 0.25% of impurities (S+P+Si+Mn), and less than 2% slag by weight. Wrought iron is used for rivets, chains, ornamental iron work, railway couplings, water and steam pipes, bolts and nuts, horseshoe bars, hand rails, straps for timber roof trusses, boiler tubes, roofing sheets etc.

2.7.2 Iron-Carbon Phase Diagram

A very large number of elements form alloys with iron to make it more useful. Carbon is the most important alloying element in this list as the properties of iron are greatly changed by the combination with carbon. Iron and carbon form an intermetallic compound namely cementite (Fe_3C), which is very hard and brittle. Fe and Fe_3C (iron carbide) are the components in the Fe–Fe_3C phase diagram (see Fig. 2.13).

However, it is also possible to have a phase diagram with Fe and C (graphite) as the components. Graphite is more stable than Fe_3C. So, the Fe–Fe_3C diagram can be considered to be a metastable phase diagram. Most steels contain only iron carbide and not graphite. Since under the equilibrium conditions the carbon in these alloys is in the form of the cementite, the system may be referred to as the iron-iron carbide system. The equilibrium diagram of iron and iron carbide is shown in Fig. 2.13 where the composition is plotted on the *x*-axis as weight % of carbon and temperature in the y-axis.

There are four different solid phases in the Fe-Fe_3C equilibrium phase diagram:

1. α–Ferrite: This phase is an interstitial solid solution of carbon in the BCC iron crystal lattice. As indicated by the Fe-Fe_3C phase diagram, carbon is the only slightly soluble alpha (α) ferrite, reaching a maximum solid solubility of 0.02 wt.% at 723°C. The solubility of carbon in α ferrite decreases to 0.005 wt.% at 0°C.

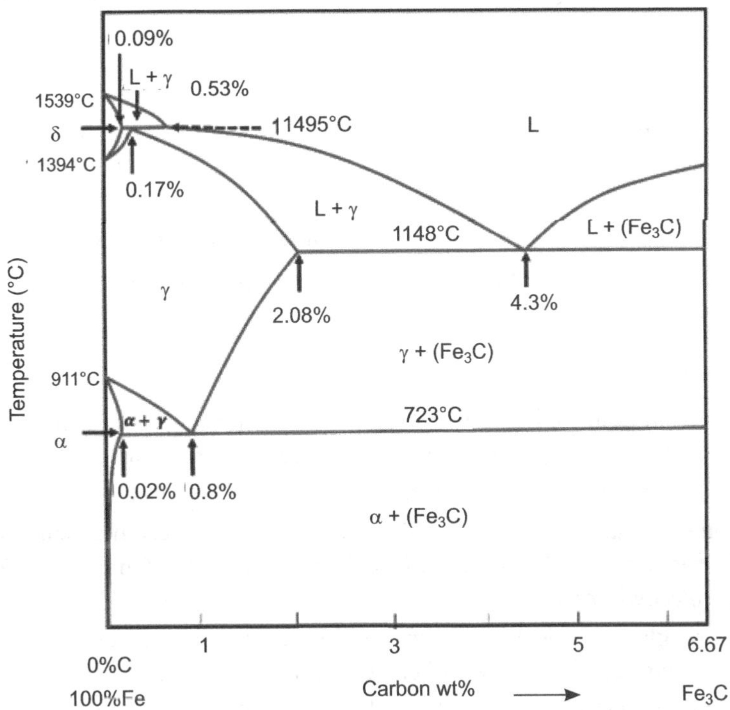

Fig 2.13: The iron-iron carbide phase diagram

2. γ–Austenite: The interstitial solid solution of carbon in gamma (γ) iron is called austenite. Austenite has an FCC crystal structure and a much higher solid solubility for carbon than α ferrite. The solid solubility of carbon in austenite is a maximum of 2.08 wt.% at 1148°C and decreases to 0.8 wt.% at 723°C.
3. Cementite: The cementite is an intermetallic compound which is denoted by Fe_3C. It has orthorhombic crystal structure. Cementite has negligible solubility limits and a composition of 6.67 wt.% C and 93.3 wt.% Fe. The Cementite is a hard and brittle phase/compound.
4. δ–Ferrite: The interstitial solid solution of carbon in δ –iron is called delta (δ) ferrite. It has a BCC crystal structure like ferrite but with a greater lattice constant. The maximum solubility of carbon in δ ferrite is 0.09 wt.% at 1495°C which is much higher than the solid solubility of carbon in α ferrite at any temperature.

2.7.3 Invariant Reactions in the Fe-Fe_3C Phase Diagram

There are three invariant reactions observed in Fe-Fe_3C equilibrium phase diagram.

Peritectic Reaction

Pure iron melts at 1539°C. A peritectic invariant reaction occurs at 1495°C with the peritectic composition at 0.17% carbon:

$$\text{Liquid } (0.53\%\text{C}) + \delta(0.09\%\text{C}) = \gamma(0.17\%\text{C})$$

δ–ferrite is a high-temperature phase and so is not encountered in plain-carbon steels at lower temperatures. On cooling through peritectic temperature, –ferrite reacts with liquid to form austenite (γ).

Eutectic Reaction

At the eutectic reaction point, liquid of 4.3% percent forms austenite of 2.08% C and the intermetallic compound Fe3C (cementite) which contains 6.67% C. This reaction, which occurs at 1148°C, can be written as

$$\text{Liquid } (4.3\%\text{ C}) = \gamma\text{-austenite } (2.08\%\text{C}) + Fe_3\text{ C } (6.67\%\text{C})$$

On cooling through eutectic temperature, liquid decomposes to austenite () and cementite (. This mixture is known as *ledeburite*.

Eutectoid Reaction

At the eutectoid reaction point, solid austenite of 0.8% C produces –ferrite with 0.02% C and Fe_3C that contains 6.67% C. This reaction, which occurs at 723°C, can be written as:

$$\gamma\text{-austenite } (0.8\%\text{C}) = \alpha\text{-ferrite}(0.02\%\text{C}) + Fe_3C\ (6.67\%\text{C})$$

A plain carbon steel that contains 0.8% C is called a eutectoid steel. If a plain-carbon steel contains less than 0.8% C, it is termed a *hypoeutectoid steel*, and if the steel contains more than 0.8% C, it is designated a *hypereutectoid steel*.

On cooling through eutectoid temperature, austenite (γ) decomposes to α–ferrite and cementite (Fe_3C). The lamella mixture, of α–ferrite and cementite, is known as *pearlite*. The interlamellar spacing is a technical term to characterize the pearlite colony. It is defined as the distance between the centre of a ferrite (or cementite) plate to the centre of the next ferrite (or cementite) plate. The amount of supercooling below eutectoid temperature and alloying elements mainly affects the interlamellar spacing.

When austenite is cooled rapidly and kept at a constant temperature, it transforms to a fine nonlamellar mixture of ferrite and cementite known as *bainite*.

All the above equilibrium phases possess different compositions, microstructures, and properties. The microstructures can be studied through microscopic techniques on prepared specimen removed from the products.

2.7.4 Cooling of Plain Carbon Steel

Cooling of austenite in a furnace yields coarse pearlite (interlamellar spacing ~ 1-2 μm) of hardness ~ 5 R_c, whereas air cooling gives fine pearlite of hardness of 20-25 R_c. Very fine pearlite (interlamellar spacing ~ 0.1 μm) of hardness 40 Rc is also possible by applying very fast cooling rate. Table 2.1 lists typical applications of steels according to carbon content.

In slowly cooled carbon steels, the overall hardness and ductility of the steel are determined by the relative proportions of the ferrite (soft & ductile) and cementite (hard & brittle). The cementite content increases with increasing carbon content, resulting in an increase of hardness and a decrease of ductility, as we go from low carbon to high carbon steels.

Table 2.1: Application of Plain Carbon Steels

Types of steel	Carbon percentage	Typical uses
Mild Steel	0.0-0.1	Car and refrigerator bodies, rivets
	0.1-0.3	Ship hulls, car bodies
	0.2-0.3	Structural components, Bridges
Medium carbon steels	0.4-0.5	Train axles and wheels
	0.5-0.6	Rails
	0.6-0.7	Chisels
High carbon steels	0.8-0.9	Saw cutters, hammers
	1.0-1.1	Axes
	1.1-1.2	Razor blades
	1.3-1.4	Scissors, Knives, files

A rivet (at the top of the list) should have good deformability, and hence has a low carbon content. In contrast, a file (at the bottom of the list) should have high hardness and wear resistance and hence has a high carbon content. Even though we may increase the hardness by increasing the carbon content further, the alloy becomes too brittle to be useful above 1.4% carbon.

Fe–C alloys with more than 2% carbon are called cast irons. Consider, for example, the cooling of a cast iron with 3% carbon, see Fig. 2.13. On crossing the liquidus into the (L+γ) region, proeutectic austenite (γ) crystallizes first. On passing through the eutectic temperature, liquid of eutectic composition decomposes to a mixture of austenite and cementite. On further cooling through the eutectoid temperature, the austenite decomposes to pearlite, yielding white cast iron. The microstructure of white cast iron consists of cementite and pearlite. Due to the presence of the large fraction of cementite, white cast iron is very hard and brittle. Depending on the cooling rate and the other alloying elements present in cast iron, the carbon may be present as graphite or cementite. Gray cast iron contains graphite in the form of flakes. Slow cooling rates and the presence of silicon promote the formation of graphite. The microstructure of gray cast iron is shown in Fig. 2.14a and consists of graphite flakes in a matrix of ferrite. The graphite flakes are sharp at their tips and act as stress raisers. Due to this, gray cast iron is brittle under tensile loads, despite the softness of graphite as compared to the very hard cementite present in white cast iron.

The brittleness can be avoided by producing the graphite in the form of spherical nodules (which do not have stress-raising sharp ends), as is done in malleable cast iron and SG (spheroidal graphite) iron (see Fig. 2.14b).

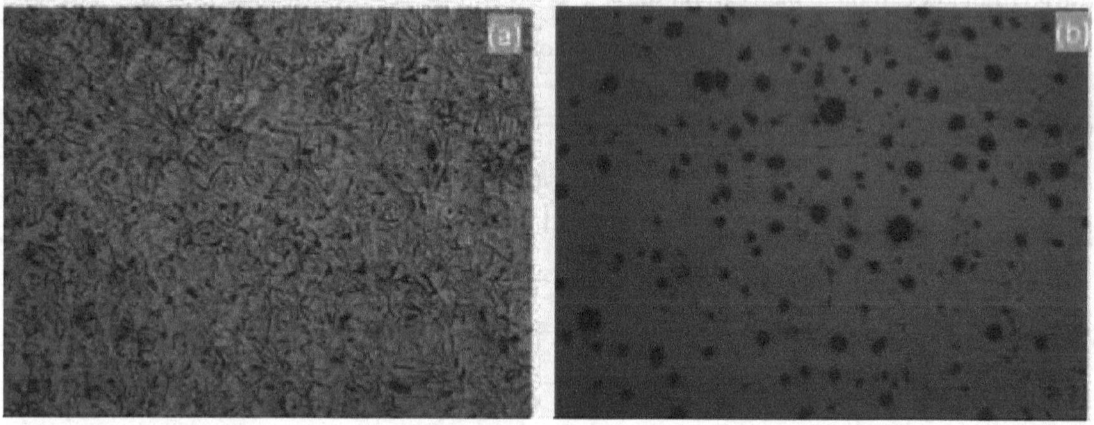

Fig. 2.14: Typical microstructure of (a) Gray cast iron, and (b) spheroidal graphite iron.

Malleable cast iron is produced by heat treating white cast iron for prolonged periods at about 900°C and then cooling it very slowly. The silicon content in the alloy must be 1% or less to ensure that cementite and not flaky graphite forms during solidification. The cementite decomposes to the more stable graphite during the subsequent heat treatment. Graphite appears in the heat-treated microstructure as approximately spherical particles of temper carbon. SG iron (also known as nodular iron) is produced by making certain alloy additions such as Mg or Ce to molten iron. Here, the silicon content must be about 2.5% to promote graphitization. The alloy additions have the effect of modifying the growth rate of graphite from the melt to be more or less equal in all directions, so that nodules (and not flakes) of graphite are produced. No subsequent heat treatment is required for SG iron.

2.8 IRON AND STEEL MAKING

2.8.1 History of Iron

The only sources of iron available to early humans were meteoric iron and native (telluric) iron. Both were scarce. Most meteorites are non-metallic; only about 6% are iron, and these contain about 7 to 15% nickel. In 1808, William Thomson sectioned and etched a meteorite, noting the remarkable patterns. Although he published his findings in 1808, they attracted little interest. Also in 1808, an Austrian, Alois von Widmannstatten, also etched a meteorite and observed the "structure¨ that is now known by his name. In 1820, A. v. Widmannstatten and Carl von Schreibers published a book on meteorites, which contained a print from a heavily etched meteorite (Figure 2.15).

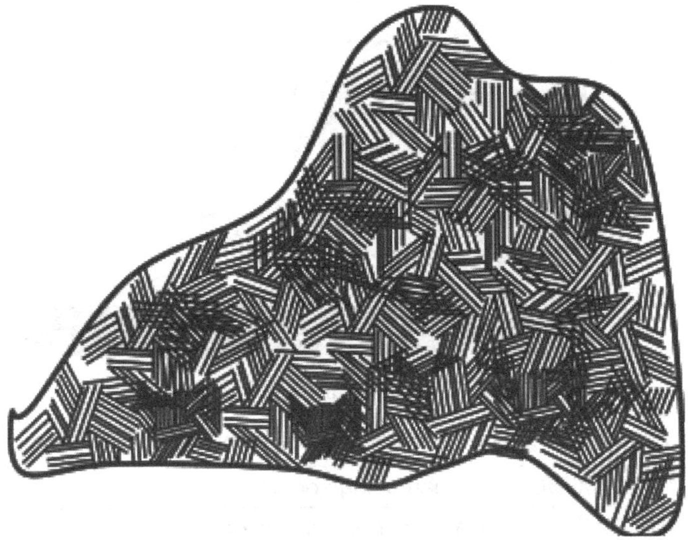

Fig. 2.15: Schematic of a heavily etched iron meteorite structure.

Native iron is even scarcer, being limited to small particles in western Greenland. Archaeological finds of iron with considerable amounts of nickel suggest that they were made from meteorites. The first production of iron dates back to at least 2000 BC in India and Sri Lanka. By 1200 BC, production of iron was widespread in China and the Near East. The most common iron ores are hematite (Fe2O3) and magnetite (Fe3O4). Smelting of iron involved heating iron ore (oxides of iron) with charcoal. The reaction of iron oxide with carbon produced carbon monoxide and carbon dioxide. The air was supplied by either a natural draft or some means of blowing. Early furnaces were of various types. An open-pit furnace is shown in Figure 2.16.

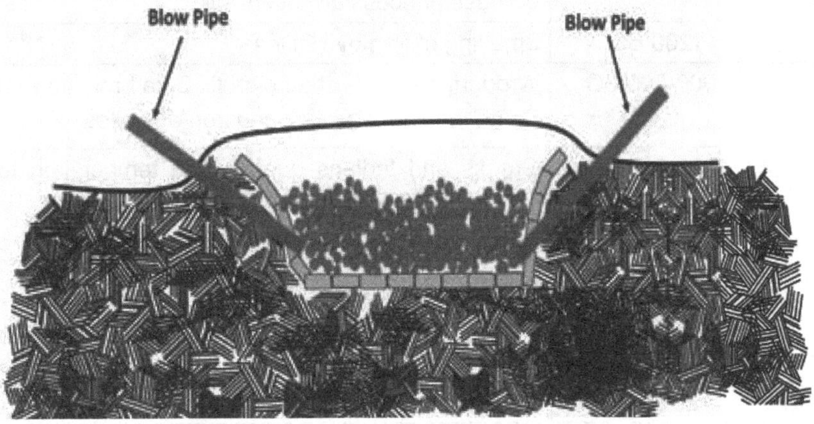

Fig. 2.16: An early open-pit iron furnace.

The carbon content of iron produced in pit furnaces was usually low because of the low temperatures achieved and resulted in semisolid sponge. With shaft furnaces (Figure 2.17), the higher temperatures resulted in higher carbon contents. In the furnaces, charcoal reacted with the air to form carbon monoxide, which reduced some of the ore.

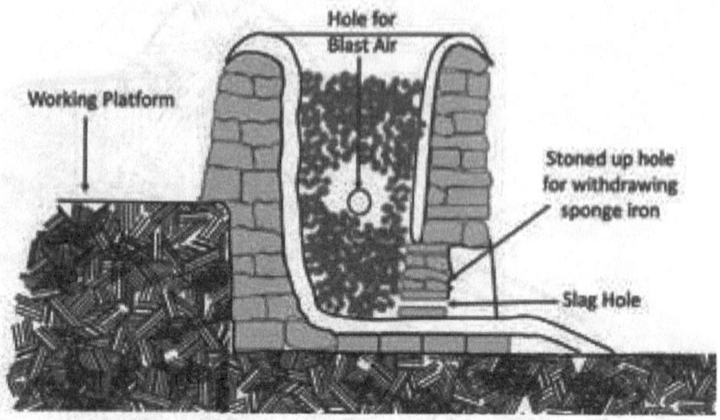

Fig. 2.17: An early shaft-type iron furnace.

Table 2.2: The three ages of civilization describing the Iron age evolution.

Stone Age	1,000,000 BC	Human beings on earth
Chalcolithic Age	8000 BC	Native copper, native gold
	6500 BC	Smelting of copper from malachite Arsenical Bronze- an accidental alloy
Bronze Age	4000 BC	Silver
	3000 BC	Tin Bronze
	2900 BC	First handmade iron object in the great pyramid of Giza
	2500 BC	Lead in Indus valley, India/Pakistan
	1750 BC	Tin
	1500 BC	Bronze by Shang dynasty in China Chinese princess discovers silk
	1200 BC	Smelting of iron by Hittites
Iron Age	1000-500 BC	Wrought and quenched high-tin beta bronze vessels in South Indian megalithic and iron age sites
	1000-700 BC	Greeks and Indians quench and temper iron to improve the cutting characteristics
	500 BC	Deepest old gold mine at Maksi, India
	500 BC	Gold, copper-gold, gold-platinum alloys
	500 BC	Reference to diamond in Indian Sanskrit texts
	300 BC	Crucible steel in South India, later known as wootz
	200 BC	Cast iron in China
	100 BC	Development of silk road
	AD400-420	Delhi Iron Pillar
	AD 1200	Zinc smelting at Zawar, India
	AD 1400	Blast furnace of iron making
	AD 1856	Bessemer steel

The resulting carbon dioxide reacted with charcoal to form more carbon monoxide. The product of the lower-temperature furnaces was low in carbon and much like wrought iron. It was soft and formable. If heated in charcoal, it would absorb more carbon and could be made into useful steel tools. The higher-temperature furnaces produced molten iron that contained up to 4% carbon. After it solidified, it formed a brittle material that was at first discarded. Later it was learned that the carbon content could be reduced by remelting in contact with air. By 200 BC, the Chinese had started casting the high-carbon material into useful objects.

2.8.2 History of Steel

Early carbon steels were made by several methods. Some were made by carburizing wrought iron without melting by having it in contact with charcoal or by melting wrought iron and adding carbon. It was also made by decreasing the carbon content of high-carbon material by oxidation. In India and Ceylon, wootz steel was produced by heating iron ore in closed crucibles with charcoal and glass. This formed small buttons of high-carbon iron (typically 1.5% C). These buttons were then consolidated by forging into ingots. The technique died out in about 1700AD when the necessary ores containing tungsten and vanadium had been depleted. The high carbon content meant that Fe3C was present, and this resulted in a very hard material with a characteristic surface pattern. Five-pound cakes of wootz steel were shipped to Persia, where they were beaten into swords. The term Damascus swords was given to the swords by Europeans, who first encountered their use by the Saracens in Damascus during the Crusades. The steel had 1.5% to 2.0% carbon, so it had large amounts of iron carbide, which made it very hard. These swords were characterized by wavy patterns resembling the surface of water, the result of bands of carbide particles, which etched white, and a steel matrix, which etched black. The bands resulted from the segregation of carbide-forming elements (principally vanadium and molybdenum) during freezing. These bands attracted carbon. Damascus swords/ blades were both resistant to fracture and capable of holding a very sharp edge probably due to the formation of carbon nanotubes (CNT) along its edge. Damascus sword were made up of wootz steel and India was technologically very advanced to produce such steels from about 1100 AD to 1700AD. European attempts to reproduce these steels by forging together layers of high and low-carbon steel were unsuccessful. It is believed that the art of producing Damascus sword involved three important factors. One was that the ore deposit contained crucial amounts of carbide-forming elements. The fact that the steels from these deposits also contained high levels of phosphorus meant that they would crack on forging unless the surface had been decarburized first. A blacksmith must also have been taught how to produce the characteristic patterns, because they would not be visible until after the decarburized surface was ground off. The secrets of making such blades were carefully guarded. Verhoeven suggested that the source of ores containing the proper impurities became unavailable, and thus the techniques stopped working. However, it is also believed that East India company people has forced to stop the production of such type of steels in India due to strategic reasons. Other ancient steel dating back to 1400 BC, comes from Africa. The people of East Africa invented a type of high-heat blast furnace that allowed them to produce carbon steel at 3275oF (1800oC) nearly 2000 years ago. This ability was not duplicated until centuries later in Europe during the Industrial Revolution. In the fourth century BC, steel weapons were produced in Spain. The earliest production of high-carbon steel was found in Ceylon (Sri Lanka). Early steel making employed the unique use of a wind furnace, blown by the monsoon winds. This made possible the production of high-carbon steel. Crucible steel, formed by slowly heating and cooling pure iron oxide and carbon (typically in the form of charcoal) in a crucible, was produced in central Asia by the 9th to 10th century AD. The development of the modern blast furnace from early shaft furnaces was a gradual process. Furnaces evolved to become large and taller. The use of coke instead of charcoal was first introduced in England in 1619 but did not become common until the early 1700s. The preheated blast was first used in England in the early 1870s.

2.8.3 Iron and Steel Fabrication Process

The earth's crust contains about 5% iron. However, it is present as a compound, mostly as iron oxide. Iron is extracted as metallic Fe from the oxide by reduction with a reducing agent. The principal agent that is used is naturally occurring carbon in the form of coke, coal or charcoal.

For economic extraction of iron, relatively rich deposits of iron ore, with Fe contents above 55–60%, are generally used. The most abundant iron oxide mineral is haematite (i.e., Fe_2O_3) followed by magnetite (i.e., Fe_3O_4). Iron ores also contain oxide minerals of some other elements such as aluminium, silicon, manganese, sulphur and phosphorus. These minerals are physically mixed with earthy materials such as sand, rock and clay. Some of these substances are harmful for the properties of the finished products. These are treated as impurities that ought to be removed during extraction.

Fig. 2.18 shows a flowchart of the integrated manufacturing process for iron and steel using the blast furnace and basic oxygen furnace (denoted BF and BOF hereinafter, respectively), which is presently the most commonly used method (51% of world steel production). After the BF-BOF process, molten steel is controlled to a target composition and temperature and is then cast by continuous casting machine to produce slabs, blooms, and billets. These castings are rolled to the required dimensions by the rolling mill to produce steel products. The smelting and refining process for iron and steel in the BF-BOF process involves the carbon reduction of iron ore (Fe_2O_3) in the BF to make molten iron, and decarburization of molten iron in the BOF to make molten steel.

Major reducing agent in the BF is the carbon monoxide gas (CO) generated by the oxidation of the carbon (C) in coke. Consequently, carburization takes place at the same time as reduction, producing hot metal (molten iron) containing about 4% carbon. The hot metal is decarburized to the required carbon content in the BOF.

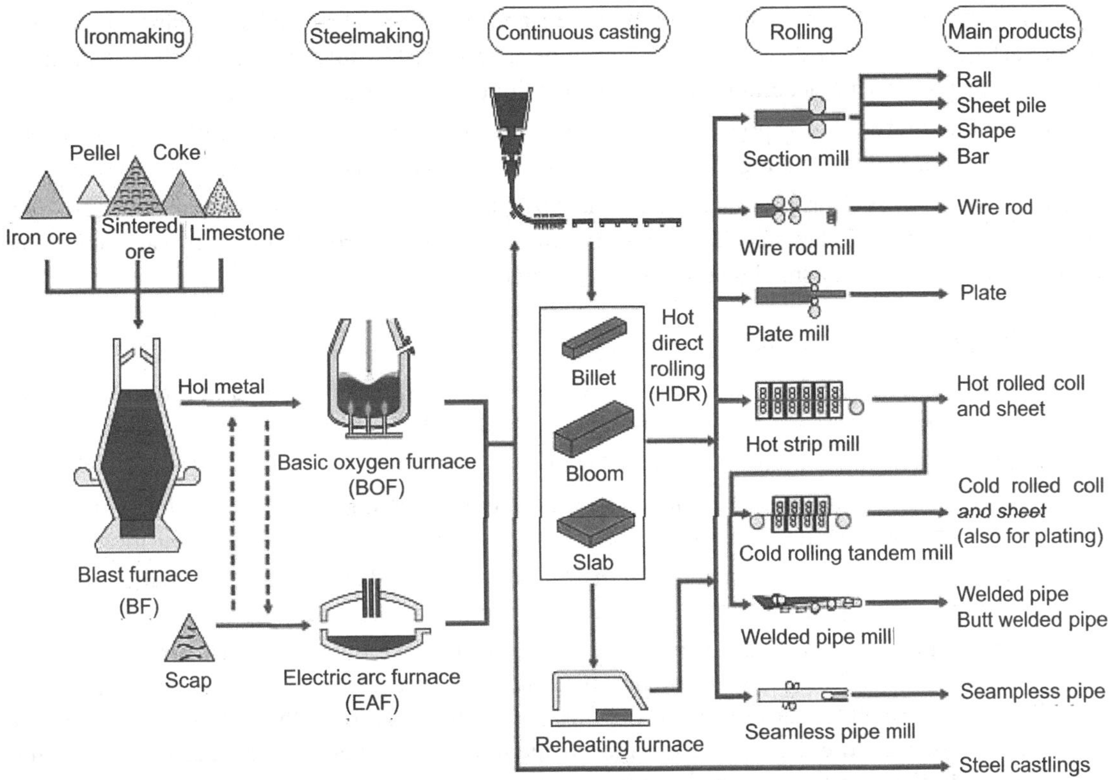

Fig. 2.18: Flow chart displaying production of Iron and Steel.

The main reaction in this process is the oxidization of the carbon in the hot metal by both pure oxygen gas (O_2) and iron oxide (Fe_2O_3). The residual oxygen, after contributing to this decarburization reaction, remains in the molten steel. This oxygen is fixed and removed by deoxidation reagents such as silicon and aluminum as SiO_2 and Al_2O^3 or is removed as carbon monoxide gas in the subsequent vacuum degassing process.

In addition to the BF-BOF process, there is another process which utilizes mainly scrap as an iron source, with some direct reduced iron whenever necessary. The direct reduced iron is produced by reducing iron ore with reformed natural gas, whose principal components are hydrogen, carbon monoxide, and methane. The scrap, along with direct reduced iron, is then melted in an electric arc furnace (EAF) to produce molten steel which is subsequently processed by the continuous casting machine, as mentioned above.

The molten steel from the BOF and EAF is then deoxidized, and alloying elements are added in the prescribed amounts. The molten steel is then held at the target temperature and continuously cast, and the castings obtained are cut to the prescribed length. After heating to the rolling temperature in a reheating furnace, these castings are hot worked to the required products. Steel shapes, bars, and wire rods are worked on section and bar mills and wire-rod mills equipped with Caliber rolls, plates are worked on reversing mills, and hot-rolled steel sheets are worked on hot strip mills. After pickling to remove scale from the surface, the hot-rolled steel sheets are worked to cold-rolled steel sheets on reversing mills or tandem rolling mills, and the cold-rolled steel sheets are tinned or galvanized as required to produce various surface-treated steel sheet products. Steel pipe is produced by forming and welding steel sheets or plates, or by piercing a billet and rolling to the final dimensions without a seam.

Iron has the ability to form alloys with many elements. Such alloying is done intentionally to produce various grades of iron alloys having the properties required for various applications.

In a broader sense, iron products can be classified into three categories, viz. pig iron, cast iron, wrought iron and steel. Today, steel is the most important product of iron, followed by cast iron, while wrought iron is used in very small quantities for ornamental purposes.

2.8.4 Pig Iron

The crude impure iron, which is extracted from iron ores, is known as pig-iron and it forms the basic material for the manufacture of cast-iron, wrought iron and steel.

The pig iron is manufactured by the following operations:

(i) Dressing: Crushed into pieces 25mm followed by removal of impurities of clay, loam and other earthy matter by washing. Further, magnetic separators are used for magnetic impurities.

(ii) Calcination and roasting: Water and carbon dioxide are removed from ores by calcinations process. By roasting, making the ores hot and very dry after removal of sulphur.

(iii) Smelting: Smelting is carried out in a special type of furnace known as blast furnace. The raw material consists of iron ores, the fluxing materials like limestone and fuel like coal, charcoal is allowed to through throat portion of the furnace. By the reduction, the pig iron collects in the hearth of furnace. The slag formed is removed and hot gases dust escapes through outlet, which is provided in the throat portion of furnace.

2.9 IRON AND STEEL HERITAGE OF INDIA

As per evidence available at present, iron objects were in use in India as early as 1100 BC. Table 2.3 represents the timeline of iron and stell history of India. During the ancient period, several communities, including women, participated in the process of iron making with a real feeling of community that is lost in the impersonal modern-day. The glorious heritage in iron and steel of India is thus unquestionable.

One example of the extent of perfection achieved by Indian artisans in yesteryears is the famous Delhi Iron Pillar. The pillar weighing nearly 7 tonnes is about 8 m in height with bottom and top diameters of 0.42 m and 0.31 m respectively. It was made about 1500–1550 years ago by forge-welding individual pieces of wrought iron. What is amazing is that it has resisted rusting for all these years. The corrosion resistance of the Delhi Iron Pillar has been the subject of investigations by many scientists taking the help of samples collected from the base of the Pillar.

Table 2.3: Timeline of Iron and Steel in Indian History

1100 BC	Iron from megalithic sites such as Hallur from Karnataka
326 BC	King Porus presented Alexander 30 pounds of Indian Iron
300 BC	Crucible processes at Kodumanal, Tamil Nadu, related to steel
300 BC	Kautilya showed knowledge of minerals, including iron ores and the art of extracting metals in Arthashastra

AD 77-79	Roman Pliny's accounts of Seric iron pointing to Chera, South India.
AD 320	A 16m iron pillar erected at Dhar (Madhya Pradesh).
AD 400-420	Gupta iron pillar (later moved to Delhi)
12th century AD	Export of iron and steel from Malabar by Jewish merchants.
13th century AD	Iron beams used in the construction of the Sun temple, Konark, Orissa.
17th century AD	Tavernier's accounts of export of 20,000 pounds of wootz steel from Golconda in Andhra Pradesh to Persia; Manufacture of cannon and firearms
AD 1830	Suspension bridge built over the Beas at Saugor with iron from Tendulkhema (MP); J. M. Heath built iron smelter at Porto Novo, Madras Presidency
AD 1870	Bengal iron works established at Kutli
AD 1907	Tata iron and steel company formed
AD 1914	Howrah bridge in Calcutta built from 80,000 tonnes of Tata steel
AD 1918	Mysore iron works at Bhadravati
AD 1954	Hindustan steel limited formed to construct three integrated steel plants at Rourkela, Durgapur and Bhilai
AD 1960	Alloy Steel Plant installed at Durgapur
AD 1972	Steel plant at Bokaro
AD 1978	Steel authority of India Limited (SAIL) formed
AD 2004	Acquisition of overseas steel plants by Tata Steel

2.10 CAST IRONS

Cast irons are the family of ferrous alloys with more than 2.08%C. As their name implies, cast irons are intended to be easily cast into the desired shape instead of being thermo-mechanically worked. The molten cast iron is led into moulds of required shapes to form what are known as cast iron castings and slag is removed from the top of cast iron at regular intervals. Cast irons are easily melted and do not form undesirable surface films when poured into mould. Cast irons usually contain 2 to 4% carbon along with 1 to 3% silicon. Cast irons can be alloyed to acquire improved hardness, wear, abrasion and corrosion resistance. Cast irons are cheaper than steel. They have comparatively low impact resistance and ductility, and this limits their wide industrial applications.

There are four types of cast irons namely white cast iron, gray cast iron, malleable cast iron, and nodular cast iron. Each of these can be distinguished from one another by differences of composition and their microstructure. Most cast irons require the presence of alloying elements other than iron and carbon to produce their characteristic microstructure.

2.10.1 White Cast Iron

The typical composition of white cast irons is 1.8–3.6% C, 0.5–1.19% Si, 0.25–0.80% Mn, 0.06–0.20 % S, and 0.06–0.18% P. The microstructure of unalloyed white cast iron contains large amount of Fe3C and pearlite. This class of alloy is called as white cast iron as the fractured surface of such alloys looks white.

White cast irons are commonly used for their excellent wear resistance.

Application of white cast iron: It is also used as raw material for malleable cast irons.

2.10.2 Gray Cast Iron

The typical composition of gray cast irons is 2.5–4.0% C, 1.0–3.0% Si, 0.25–1.00% Mn, 0.02–0.25% S, and 0.05–1.0% P. The microstructure of unalloyed gray cast iron contains graphite flakes in a matrix of mixture of ferrite and pearlite (see Fig. 2.14a). However, fully ferritic matrix is possible after suitable heat treatment. This class of alloy is called as gray cast iron as the fractured surface of such alloys looks gray. Gray cast iron is an important engineering material because of its low cost and excellent properties such as vibrational damping capacity, machinability, and wear resistance. The properties of gray cast iron depend on the morphology and size of the graphite flakes and the matrix. There are four types of morphology of graphite flakes as mentioned in AFA-ASTM standard. The most desirable microstructure is fine and randomly oriented (type A) flakes.

Application of gray cast iron: Locomotive and internal combustion engine cylinder blocks and heads, flywheels, counterweights for lifts, etc.

2.10.3 Malleable Cast Irons

The typical composition of malleable cast irons is 2.0–2.6% C, 1.10–1.6% Si, 0.20–1.00% Mn, 0.04–0.18% S, and 0.18% (max.) P. The microstructure of unalloyed malleable cast iron contains spheroidal carbon particles (temper carbon) in a matrix of ferrite. Malleable cast iron is an important engineering material because of its high tensile strength (700 MPa max.) and better elongation (10-15%). They are relatively costly due to heat treatment involved in converting white cast iron into malleable cast iron.

Application of malleable cast iron: Automobile crankshaft, chain links, brackets, etc.

2.10.4 Nodular Cast Irons

The typical composition of nodular cast irons (also called as spheroidal graphite irons or ductile irons) is 3.0–4.0% C, 1.8–2.8% Si, 0.10–1.00% Mn, 0.03% (max.) S, and 0.10% (max.) P. Magnesium is used a modifier. Magnesium reacts with sulphur and oxygen so that these elements cannot interfere with the formation of the nodular graphite (see Fig. 2.14b). The microstructure of unalloyed nodular cast iron is usually of the bull's eye type structure. This structure consists of "spherical" graphite nodules with envelopes of free ferrite around them in a matrix of pearlite. Nodular cast iron is an important engineering material because of its high tensile strength (400–700 MPa) and better elongation (10-18%).

Application of nodular cast iron: Crankshaft, agricultural implements, coke oven doors, gears, industrial fan hub, etc.

2.11 STEEL

Steel is mainly a combination of two elements, iron and carbon. A wide variety of steels are in common use and many of them are sold under trade names. Steel is classified into four types according to: (1) types of production (Basic open-hearth, electric, basic oxygen process, etc.), (2) forms and use (Deep drawing steels, boiler steels, electrical steels, free-cutting steel, spring steels, etc.), (3) grade (Plain carbon steel, alloy steel, etc.), (4) quality. In fact, there are different types of steels, and each type of steels has unique characteristics which lead it to being used in its respected fields.

2.11.1 Steel Specifications

Steel specification is required for engineers to select the right steel for a specific application. Steel specification vary from country to country. Chemical composition, mechanical properties, or hardenability are the criteria used to specify steels. Most specifications are based on chemical composition of steel.

Indian standard specifications

The Bureau of Indian Standards (BIS) has published several standards related to classification of steels. The standard IS 7598: 1990 (IS stands for Indian Standard) deals with the classification of steels based on chemical composition but provision is also made for subdivision according to specific characteristics or use. The standard IS 1956 (Part 2): 2018 deals with definitions of the terms commonly used in the field of steel making and classification of steel. The concentration of carbon and other alloying elements are taken into consideration in the classification of steels. For example, IS: C15 is a carbon steel with 0.15 wt.%C. Similarly, 20Mn1 stands for 0.20 wt.% C, 1wt.% Mn, and 20Mn60 stands for 0.20 wt.% C, 0.60wt.% Mn. The underlining of number (i.e. 60) indicates that manganese content is 0.60 wt.%. The same notation is valid for low alloy steels. However, the letter 'T' is used at the start in BIS to denote tool steels. For example, IS: T105Cr1Mn60 is a tool steel with 1.05 wt.%C, 1 wt.% Cr, 0.60 wt.% Mn.

AISI-SAE specifications

India is performing well in manufacturing business. As a technical person may have to deal with export/import of steels. Therefore, knowledge about foreign specifications is essential and useful. In AISI-SAE (AISI stands for American Iron and Steel Institute, and SAE stands for Society for Automotive Engineers) standard, the composition is denoted by a four-digit number. The first two digits indicate the principal alloying element or group of elements in the steel, and the last two digits indicate the hundredths of percent of carbon in the steel. For example, AISI 1040 stands for plain carbon steel which is equivalent to IS: C40. AISI 4340 stands for Ni-Cr-Mo steel, with 1.83%Ni, 0.5 or 0.8%Cr, 0.25%Mo, and 0.4%C. Table 2.4 lists the meaning of these digits for some of the standard steels.

Specifications of other countries

The old British specifications for steels have the letters En (En stands for emergency number) followed by a number. For example: En 8 is a plain carbon steel with 0.4% C which is equivalent to AISI 1040 and IS: C40. In the new British specification BS970, the first three digits stands for the alloy content. The middle letter A, M, or H stands for analysis, mechanical properties, or hardenability respectively. The last two digits indicate the carbon content. The German specifications starts with the letters DIN, the Russian with GOST, and the Japanese with JIS.

2.11.2 Plain Carbon Steel

Carbon steels are the most common types of steel accounting for around 90% of steel production. Carbon steel typically contains up to 2 wt.% of carbon. Carbon steel is classified into three main subcategories, low carbon steel, also known as mild steel, medium carbon steel, and high carbon steel. The term carbon steel can also be used as a catch call term for steel that isn't stainless steel. There are three different types of Carbon Steels namely low carbon steel, medium carbon steel, and high carbon steel.

Low Carbon steel

Low carbon steel (or "mild carbon steel" or "plain carbon steel") refers to carbon steels that have up to 0.30% carbon content. This is, by far, the most common type of steel on the metals market. There are a few reasons for this. For starters, low carbon steel is relatively economical. Also, since the carbon content is lower than medium and high steels, low carbon steel is easy to form and is perfect for applications where tensile strength isn't an immediate issue, such as structural beams. Another benefit of low carbon steel is that its properties can be improved relatively easily by adding additional elements, such as manganese. Low carbon steel is also an ideal choice for carburizing, which improves case hardness without impacting ductility or toughness.

Applications of low carbon steels: Structural components, machinery, pipes, domestic appliances, automobile components, surgical instruments, medical equipment, wires, bolts, stampings, etc.

Table 2.4: AISI-SAE classification of steels

AISI Classification	Types of steel
10xx	Plain carbon steel
11xx	Resulphurized grades
13xx	Manganese steels (Mn: 1.6-1.9%)
40xx	Mo steels (Mo: 0.20% or 0.25%)
41xx	Cr-Mo steels (Cr: 0.50%,0.80%, or 0.95%; Mo: 0.12%,0.20%,or 0.30%)
43xx	Ni-Cr-Mo steels (Ni: 1.83%; Cr: 0.50%, or 0.80%; Mo: 0.25%)
51xx	Chromium steels (Cr: 0.80%, 0.88%, 0.93%, 0.95%, or 1.00%)
51xxx	Medium Chromium bearing steels (Cr: 1.03%)
52xx	1.45%Cr steels
61xx	Cr-V steels (Cr: 0.60%, or 0.95%; V: 0.13% or 0.15%)
86xx	Ni-Cr-Mo steels (Ni: 0.55%; Cr: 0.50%; Mo: 0.20%)
92xx	Si-Mn steels (Si: 1.4% & Cr: 0.7% , or Si: up to 2.0%; Mn: 0.60–0.90 %)
50Bxx	Boron treated chromium steels (Cr: 0.50%; C: 0.60%)

Medium Carbon Steel

Medium carbon steel refers to carbon steels that have between 0.31% and 0.85% carbon content and between 0.31% and 1.60% manganese. One of the biggest benefits of medium carbon steel is its strength. However, that comes with some trade-offs. Medium carbon steel has low ductility and toughness — making it more difficult to form and weld.

Applications of medium carbon steels: Machinery parts, pressure structures, crank shafts, gears, connecting rods, railway tracks, etc.

High Carbon Steel:

High carbon steel refers to carbon steels that have between 0.85% and 1.50% carbon content and between 0.31 and 0.90 manganese. When it comes to hardness and toughness, high carbon steel is the carbon steel of choice. However, this comes at a trade-off. It's very difficult to weld, cut, or form high carbon steel.

Applications of high carbon steels: leaf or coil springs, dies, wire and cables, saws, razors, files, jewellers' files, etc.

Plain carbon steels are employed successfully where strength and other requirements are not stringent. The common limitations of plain carbon steels are as follows:

- Low strength at elevated temperature.
- Low corrosion and oxidation resistance.
- Sharp reduction of hardness on tempering.
- Low hardenability (Hardenability of steel is the ability to be hardened).
- Low resistance to abrasion/wear.

2.12 ALLOY STEELS

The widest and most diverse range of steel alloys is "alloy steels". Alloy steels are classified in Table 2.3. According to IS7598: 1990, alloy steels can be subdivided, based on total alloying elements, into three categories namely low alloy steel (Up to and including 5%), medium alloy steel (More than 5% but up to and including 10%), and high alloy steel (more than 10%). These are made by combining carbon steel with various alloying elements to provide unique properties to each steel. Alloy Steels use different types of elements to influence the characteristics of the metal. Common elements are chromium, nickel, copper, molybdenum, vanadium, and aluminium.

2.12.1 Purpose of Alloying

Alloying elements are added to Fe–C system to overcome the above limitations and to further enhance the outstanding characteristics of plain carbon steels. Further, heat treatment of alloy steels can provide a wide range of beneficial qualities including enhanced corrosion resistance, increased hardenability, superior strength, and hardness. The possible effect of adding alloying elements may be summarised as follows:

- It may form solid solution, intermediate compounds, or intermetallic compounds.
- It may change the phase transformation temperatures of invariant reactions.
- It may alter the solubility limit of carbon in alpha, gamma, and delta iron.
- It may alter the kinetics of phase transformations occurring through invariant reactions.
- It may decrease the softening effect on tempering.
- It may provide greater depth of hardenability in larger sections.

2.12.2 Silicon Steel

Silicon steel is a soft magnetic material that is used in electrical power transformer, motor, and generator. The addition of up to 6.5 wt% silicon yielded the best electrical and magnetic properties. However, the material becomes brittle and difficult to work with using the traditional manufacturing methods when Si is above 3.2 wt%. Therefore, silicon content of about 3.2 wt.% is widely used as soft magnetic materials which has desired electrical resistivity and, therefore, reduces eddy current losses. The carbon content is maintained below 0.01% to minimize precipitation and interstitial effects on crystal lattice, which tend to increase the hysteresis loss and reduces the magnetic permeability. In summary, silicon has following effects on Fe-C system:

- Increases the electrical resistivity of iron, thus reducing eddy current effects with alternating current.
- Favours high-density permeability (for definite grain orientation).
- Decrease the solubility of carbon in iron.
- Increase the grain size.
- Increases brittleness.

Silicon steel is undoubtedly the most important soft magnetic material in use today. The Silicon-manganese steels, such as AISI 9255and AISI 9260 (Si: 1.8–2.2%, Mn: 0.7–1.0) have been used extensively for leaf springs due to their high elastic ratio. However, the shock resistance of 92xx steels is inferior to that of Ni-Cr-Mo/Cr-V/C-Mo steels.

Applications of silicon steels: Transformer and generator laminators, leaf springs, etc.

2.12.3 High-Speed Steel

High-speed steel (HSS or HS) is a subset of tool steels, which is commonly used as cutting tool material. High speed steels (HSS) get their name from the fact that they may be operated as cutting tools at much higher cutting speeds than is possible with plain carbon tool steels. High-speed steels operate at cutting speed 2 to 3 times higher than for carbon steels. In summary, high-speed steels have the following characteristics:

- Able to retain hardness at elevated temperature.
- Deep-hardening capability.
- Wear and abrasion resistance.

Chromium is one of the principal elements that controls hardenability in these steels. Vanadium increases the cutting ability of high-speed steels. Cobalt increases the cutting ability of high-speed steels at the cost of hardenability. A common type of high-speed steel contains 18% tungsten, 4% chromium, 1% vanadium, and only 0.5–0.8% carbon. It is an alloy tool steel that retains much of its hardness and toughness at red heat after heat treatment, allowing tools made of it to cut at high speeds even while red-hot through friction. When a hard material is machined at high speed with heavy cuts, sufficient heat may be developed to cause the temperature of the cutting edge to reach a red heat. This temperature would soften carbon tool steel containing even up to 1.5 per cent carbon to the extent of destroying their cutting ability. Certain highly alloyed steels, designated as high-speed steels, therefore, have been developed which must retain their cutting properties at temperatures up 600°C to 620°C.

The replacement of tungsten with molybdenum in high-speed steel reduces its cost. They are more difficult to heat treat as they are more susceptible to oxidation at quenching temperature. The high-speed steels maintain above characteristics due to presence of large amount of tungsten and/or molybdenum, together with chromium, cobalt, or vanadium. The high-speed steels are either tungsten-based types (Type T) or molybdenum-base types (Type M). The hardness and toughness of these steels are controlled primarily by the carbon content. Alloying elements (W, Mo, Cr, V) present in these steels are responsible to obtain high hardness (up to 60-67 HRC). The cobalt addition to high-speed steels permits cutting hard, gritty, or scaly materials.

Applications of high-speed steels: High cutting tools, blanking dies, hot-forming dies, lathe centers, wearing plates, etc.

2.12.4 Heat Resisting Steel

Heat resisting steels can resist tempering and have adequate creep resistance at high temperature. The Fe-Cr ferritic and Fe-Cr-Ni austenitic stainless steels are considered as heat resisting steels. The 18-8 types of austenitic stainless steel maintain satisfactory strength up to 700oC. The 20-10, 29-10, 25-12, 28-15, and 25-20 percent Cr-Ni austenitic stainless steels have greater strength at elevated temperatures and can withstand thermal fluctuations. Stainless steels in which nickel exceeds the chromium content are used where thermal shocks are present.

Applications of heat resisting steels: Heat treatment furnace parts, quenching fixtures, etc.

2.12.5 Spring Steels

Spring steels (0.50-0.65%C) are quenched and tempered to attain high yield strength (up to 1500 MPa). Spring steel is a name given to a wide range of steels *used in the manufacture of different products*, including swords, saw blades, springs and many more. These steels are generally low-alloy manganese, medium-carbon steel, or high-carbon steel with a very high yield strength. This allows objects made of spring steel to return to their original shape despite significant deflection or twisting (high resilience). The Si-Mn & Cr-V steels are commonly used as spring steels. Example: 55Si2Mn90, and 50Cr1V23.

Proper heat treatment of spring steels is required to be performed to attain fine grain size and low residual stress which will ensure their high fatigue strength.

2.12.6 Magnet Steel

Lodestone was the earliest natural magnetic materials. Later heat-treated high carbon steel magnets were used for centuries as compass needles. By 1855 a steel containing about 5% tungsten was in use for magnets. By 1917, Japanese investigators reported that a steel containing 30-40% cobalt, plus tungsten and chromium had a coercivity of 230 Oe. This is the best magnet steel.

Iron based permanent magnets were also investigated. Cunife (60% Cu, 20% Ni, 20% Fe) and Remalloy or Comol (68% Fe, 17% Mo, 12% Co) are permanent ductile magnetic alloy which are not commonly used now. There are other iron based magnetic steels namely Vicalloy 2 (52% Co, 38% Fe, 10% V), Chromindur (61.5% Fe, 28% Cr, 10.5% Co), Arnokrome (26-30% Cr, 7-10% Co, balance Fe), etc. It is also true that these magnetic steels show inferior magnetic properties as compared to rare earth material based permanent magnets.

2.12.7 Stainless Steel

Stainless steels are generally selected for service conditions where corrosion and oxidation resistance are crucial. To make 'stainless steel' stainless, there must be at least 12% chromium in the steel as chromium forms a protective oxide layer at the surface of these steels. as engineering materials Stainless steel's corrosion resistance and mechanical properties can be further enhanced by adding other elements, such as nickel, molybdenum, titanium, niobium, manganese, etc.

There are four types of stainless steels: ferritic, martensitic, austenitic, and precipitation-hardening.

Ferritic stainless steel

Ferritic stainless steels are essentially Fe-Cr binary alloys containing 12 to 30% Cr. When (%Cr – 17 X %C) > 12.7, the stainless steel is in ferrite state and their microstructure remains ferritic at normal conditions. They do not respond to heat treatment. These steels possess excellent corrosion and oxidation resistance. Examples: AISI 430 (16% Cr and 0.12%C) & AISI 446 (25% Cr and 0.20%C). The yield strength, tensile strength, and elongation of AISI 430 are 350 MPa, 550 MPA, and 30% respectively.

Application of ferritic stainless steel: Mainly for Construction work.

Martensitic stainless steel

Martensitic stainless steels are essentially Fe-Cr binary alloys containing 12 to 17% Cr with sufficient carbon (0.15–1.0%) so that martensitic structure can be produced by quenching the steel from austenitic region. When (%Cr – 17 X %C) < 12.7, the stainless steel can be martensitic stainless steel. They respond to heat treatment. These steels have comparatively poor corrosion resistance as compared to ferritic and austenitic stainless steels. Examples: AISI 410, AISI 416, and AISI 440C. The yield strength, tensile strength, and elongation of AISI 410 are 1200 MPa, 1300 MPA, and 5% respectively.

Application of martensitic stainless steel: Machine parts, bolts, bushings, pump shaft, etc.

Austenitic stainless steel

Austenitic stainless steels are essentially Fe–Cr–Ni ternary alloys containing 16 to 25% Cr and 7 to 20% Ni. Austenitic stainless steels have comparatively better corrosion resistance than that of ferritic and martensitic stainless steels. Examples: AISI 301, AISI 304, AISI 316, and AISI 347. AISI 304 (18% Cr, 8% Ni, and 0.08% C max.) is the most common types of 300 series of austenitic stainless steel. The yield strength, tensile strength, and elongation of AISI 304 are 240 MPa, 600 MPA, and 65% respectively. However, the yield strength of austenitic stainless steel can be increased up to 1000 MPa. They are susceptible to intergranular corrosion.

Application of austenitic stainless steel: Chemical industry, surgical implants, household, and sanitary fitting, etc.

Precipitation-hardening stainless steel

The precipitation hardening (PH) stainless steels or controlled-transformation steels are low carbon martensitic, semi-austenitic and austenitic stainless steels. They respond to heat treatment. The yield strength and tensile strength of PH stainless steels can be 520–1500 MPa and 850–1700 MPa. They have excellent corrosion resistance and better mechanical strength as compared to ferritic stainless steels. Precipitation hardening is achieved by the addition of aluminium, titanium, copper, and molybdenum either singly or in combination. Examples: 17/4PH (16.50%Cr, 4.25%Ni, 4.25%Cu, 0.75%Mn, 0.30%Nb, 0.05%C; Structure: martensitic), 17/7PH (17.25%Cr, 7.25%Ni, 1.25%Cu, 0.70%Mn, 0.06%C; Structure: austenitic–martensitic), PH15/7Mo (15.25%Cr, 26.00%Ni, 1.25%Mo, 2.15%Ti, 1.45%Mn, 0.25%V, 0.007%B, 0.04%C; Structure: austenitic).

Application of austenitic stainless steel: oil and gas, nuclear and aerospace industries.

UNIT SUMMARY

1. A phase diagram is basically the graphical representation of different phases present in the materials system at various pressures, temperatures, and compositions.
2. Phase diagrams discussed in this unit are experimentally constructed using the information collected from cooling curves at constant pressure.
3. The phase rule gives the degree of freedom that are permissible (i.e. number of independent thermodynamic parameters) for a system of a given number of components and number of phases in equilibrium.
4. The temperature-composition diagram of Fe-Fe3C has three invariant reactions namely peritectic, eutectic, and eutectoid reactions.
5. Commercial alloys are generally located in the simpler parts of the phase diagrams.
6. One can gather mainly three types of information from phase diagrams at a particular temperature: (i) what phases exists, (ii) what are the phase compositions, and (iii) what are the fraction of phases.
7. Indian iron technology was exemplary. Delhi iron pillar and wootz steel from India are considered as an advanced ferrous material.
8. The steels are the most important alloys, mainly because of wide range of mechanical properties and low cost. A detailed classification of plain-carbon steel and alloy steels have been discussed in this unit.
9. The elements carbon and manganese are the most important ingredients in steel. Other alloying elements are also added to further tailor the properties of steels.
10. All four types of cast irons are industrially important class of ferrous alloys.
11. Stainless steels are mainly selected for service conditions where corrosion and oxidation resistance are crucial. To make a stainless steel "stainless", it must have at least 12% Cr.
12. In this chapter we have discussed several other ferrous alloys and their applications.

EXERCISES

Multiple Choice Questions (Note: some questions may have more than one correct answer)

1. Which of the following systems belongs to the substitutional solid solution category?

(a) Fe–Cr (b) Mg–Al
(c) Fe–C (d) Cu–Ni

2. Which of the following systems belongs to the interstitial solid solution category?

(a) Mg–H (b) Al–Si
(c) Fe–C (d) Fe–N

3. Which of the following equation represents the Gibbs phase rule?

(a) $P + F = C - 2$ (b) $P + F = C + 2$
(c) $P + F = C - 1$ (d) $P + F = C$

4. Which of the following equation represents the Gibbs phase rule for condensed material?

(a) P + F = C + 1
(b) P + F = C + 2
(c) P + F = C – 1
(d) P + F = C

5. Number of degrees of freedom at a triple point in a single component (unary) system is?
(a) 1
(b) 2
(c) 3
(d) 0

6. The phase boundary between a liquid phase with liquid and solid phase mixture is -------.
(a) Solidus line
(b) Liquidus line
(c) Solvus line
(d) Tie line

7. The line joining a solid phase with liquid and solid phase mixture is known as
(a) Solidus
(b) Liquidus
(c) Solvus line
(d) Tie line

8. The line joining two different solid phases is known as
(a) Solidus
(b) Liquidus
(c) Solvus line
(d) Tie line

9. In a unary system, the maximum number of phases that can coexist in equilibrium is
(a) 1
(b) 2
(c) 3
(d) 4

10. In a binary system, the maximum number of phases that can coexist in equilibrium is
(a) 1
(b) 2
(c) 3
(d) 4

11. Unary phase diagrams are ____________
(a) Temperature–Composition plots;
(b) Temperature–Pressure plot
(c) Pressure–Composition plots;
(d) All of these

12. Binary phase diagrams are most commonly ____________
(a) Temperature–Composition plots;
(b) Temperature–Pressure plots;
(c) Pressure–Composition plots;
(d) All of these

13. Which of the following systems have three-phase equilibrium at fixed temperature
(a) Eutectic;
(b) Peritectic;
(c) Isomorphous;
(d) Eutectoid

14. Lever rule is a simple rule to determine the relative amounts of two phases for a given composition of an alloy. This rule is based on
(a) Mass balance;
(b) Volume balance;
(c) Area balance;
(d) All

15. Which of the following represents a group of metals having HCP crystal structure?
(a) Mg, Be, Zn
(b) Mg, B, Ti
(c) Ti, Cd, Co
(d) Zn, Fe, Ni

16. What is the crystal structure of pure iron between 1000°C and 1350°C?
(a) HCP
(b) BCC
(c) FCC
(d) BCC and FCC

17. Which of the following is a thermodynamically stable defect?
    (a) Dislocation (b) vacancy
    (c) crack (d) precipitate
18. Which crystal system has the lowest symmetry among the followings?
    (a) Trigonal (b) cubic
    (c) hexagonal (d) Triclinic
19. Which of the following steels were used to make Tipu Sultan's sword?
    (a) High carbon steels; (b) Low carbon steels;
    (c) Highly alloyed steels
20. The concentration of Mn in IS: 15Mn75 is?
    (a) 15 wt.%; (b) 75 wt.%;
    (c) 0.75 wt.%; (d) 0.15 wt.%
21. Which of the following famous steels were used to make Damascus sword?
    (a) Stainless steels; (b) Maraging steels;
    (c) Wootz steels; (d) All
22. The concentration of carbon in IS: C07 is?
    (a) 0.07 wt.%; (b) 7 wt.%;
    (c) 1 wt.%; (d) None of these
23. As per AISI-SAE standard, the plain carbon steel is denoted as?
    (a) 10xx; (b) 13xx;
    (c) 14xx; (d) 43xx
24. As per AISI-SAE standard, the Ni-Cr-Mo steel is denoted as?
    (a) 10xx; (b) 43xx;
    (c) 13xx; (d) 40xx
25. As per AISI-SAE standard, the boron treated steel having 0.40 wt.%Cr, 0.6 wt.%C is denoted as?
    (a) 6040; (b) 60B40;
    (c) 40B60; (d) 4060

ANSWERS															
1.	(b)	2.	(b)	3.	(b)	4.	(a)	5.	(d)	6.	(b)	7.	(c)	8.	(c)
9.	(b)	10.	(c)	11.	(c)	12.	(a)	13.	(a)	14.	(a)	15.	(a)	16.	(c)
17.	(b)	18.	(d)	19.	(a)	20.	(c)	21.	(c)	22.	(a)	23.	(a)	24.	(b)
25.	(c)														

Short and Long Answer Type Questions

1. Differentiate between metal and an alloy with an example.
2. Differentiate between component and phase with an example.
3. Define a system and what determines if a system is in a state of equilibrium?
4. State Gibbs' phase rule. What is the physical significance of this rule?
5. State Lever rule. What is the physical significance of this rule?

6. Name the three allotropes of iron and state the lattice structure of each along with transformation temperature.
7. Draw cooling curves of pure iron and a eutectoid steel.
8. What are the characteristics of commercially pure iron? Write down some of its uses.
9. Draw the equilibrium diagram of the Fe-Fe3C system. Write down the three invariant reactions that takes place in this diagram.
10. Distinguish between the following three types of plain carbon steels:
11. What does the pearlite microstructure consist of? What is the microstructural difference between commercially pure iron and pearlite?
12. What does the bainite microstructure consists of? What is the microstructural difference between bainite and pearlite?
13. Describe a basic oxygen process for converting pig iron into steel.
14. Indicate the chemical composition of steels as per the BIS: 65Mn1, 25Mn75, 25Mn1S15, 55Si2Mn90, 50Cr1V23, 40Ni2Cr1Mo25, T118, T105Cr1Mn60, and T75W16Cr4V1.
15. What are the major limitations of plain carbon steels?
16. What are the purposes of using alloying elements in steels?
17. Why wootz steel was so special and is considered as an advanced material?
18. Write a short note on the contribution of Indian metallurgy in iron and steel sector.
19. Discuss about the types of cast iron and use clear schematic to describe their microstructure features.
20. Which of the ferrous materials would you like to suggest for a high temperature application and why?
21. Design a Cu-Ni alloy that will exist at a completely molten state at 1300°C and become completely solid at 1200°C.
22. Design a steel that will have 80% ferrite and 20% cementite at room temperature.

Numerical Problems

1. Derive the Lever rule for a case shown in Fig. 2.10.
2. Consider an alloy containing 30%Cu and 70%Ni. Consider an equilibrium condition and refer Fig. 2.8 to answer the followings:
   (a) How many degrees of freedom are there at 1300°C?
   (b) What phases are present at 1300°C?
   (c) What is the chemical composition of each phase at 1300°C?
   (d) What amount of each phase is present at 1300°C?
   (e) Sketch the microstructure of the alloy at 1300°C by using the circular microscopic field.
   (f) Sketch the microstructure of the alloy at room temperature by using the circular microscopic field (assume equilibrium cooling).
3. Consider an alloy containing 30%Cu and 70%Ni. Consider an equilibrium condition and refer Fig. 2.8 to answer the followings:
   (a) How many degrees of freedom are there at 1200°C?
   (b) What phases are present at 1200°C?

(c) What is the chemical composition of each phase at 1200°C?

(d) What amount of each phase is present at 1200°C?

(e) Sketch the microstructure of the alloy at 1200°C by using the circular microscopic field.

(f) If this alloy is cooled from 1300°C. Then sketch the microstructure of the alloy at room temperature by using the circular microscopic field (assume nonequilibrium cooling).

4. A eutectoid plain-carbon steel is slowly cooled from 800°C to 721°C and hold at that temperature so that the austenite is completely transformed to ferrite and cementite. Calculate the wt.% of ferrite and cementite formed.

5. A eutectoid plain-carbon steel is slowly cooled from 800°C to room temperature.
   (a) Sketch the microstructure of the alloy at room temperature by using the circular microscopic field (assume equilibrium cooling).
   (b) Sketch the microstructure of the alloy at room temperature by using the circular microscopic field (assume nonequilibrium cooling).

6. A 1.2 wt.% hypereutectoid plain carbon steel is slowly cooled from 950°C to a temperature just above the eutectoid temperature i.e. 723°C.
   (a) Calculate the wt.% of proeutectoid cementite and austenite present in the steel.
   (b) Sketch the microstructure of the alloy at a temperature just slightly above 723°C by using the circular microscopic field (assume equilibrium cooling).

7. A 1.2 wt.% hypereutectoid plain carbon steel is slowly cooled from 950°C to a temperature just below the eutectoid temperature i.e. 723°C.
   (a) Calculate the wt.% of cementite and ferrite present in the steel.
   (b) Calculate the wt.% of proeutectoid cementite, eutectoid cementite and ferrite present in the steel.
   (c) Sketch the microstructure of the alloy at a temperature just slightly below 723°C by using the circular microscopic field (assume equilibrium cooling).

8. A plain carbon steel contains 40 wt.% proeutectoid ferrites. What is its average carbon content?

PRACTICAL

Experiment title: Study the microstructure of grey cast iron.

Apparatus/ materials used: Grey cast iron sample, emery papers, alumina powder, optical microscope.

Objectives:

(a) Observe the microstructure of grey cast iron using optical micrograph and identify the phases present.

(b) determine the volume fraction of the phases present in the microstructure.

Theory:

Students can refer to ASTM 247-19 to evaluate the microstructure and distribution of graphite in grey cast iron structures, and ASTM E562 to understand the test method for determination of volume fraction in the microstructure of a given metal using manual point count method.

Procedure:

Students can refer to ASTM E3-11 for standard guide for preparation of metallographic sample preparation.

Result:

The student should show the microstructure of the grey cast iron obtained in the optical microscope along with the scale and magnification used.

Conclusion:

The phases present in the microstructure of grey cast iron are ....

The volume fraction of the different phases is ....

KNOW MORE

Students can refer to different heat treatment methods for steel.

Advancement in green steel production.

REFERENCES AND SUGGESTED READINGS

1. V. Raghavan, Physical Metallurgy Principles and Practice, PHI Learning Private Limited, Second Edition 2009.
2. Donald S. Clark and Wilbur R. Varney, Physical Metallurgy for Engineers, CBS Publishers & Distributors Pvt. Ltd., 2004.
3. William F. Smith and Javad Hashemi, Foundations of Materials Science and Engineering, Mac Graw Hill, Fourth edition, International Edition 2006.
4. Y. Lakhtin, Engineering Physical Metallurgy, CBS Publishers & Distributors Pvt. Ltd., 1998.

Dynamic QR code for further reading

3 Non-Ferrous Metals and Their Alloys

UNIT SPECIFICS

Through this unit the following aspects of non-ferrous (non-iron based) metals and their alloys have been discussed:

- Properties and uses of non-ferrous metals: aluminium, copper, tin, lead, zinc, magnesium and nickel;
- Aluminium alloys: Duralumin, Hindalium, Magnalium– composition, properties and uses;
- Copper alloys: Brasses, Bronzes – composition, properties and uses;
- Magnesium alloys;
- Nickel alloys: Inconel, Monel, Nichrome – composition, properties and uses;
- Titanium alloys;
- Anti-friction/Bearing alloys: Various types of bearing bronzes - Standard commercial grades as per BIS/ASME.

The practical applications of the non-ferrous metals and alloys are discussed for generating further curiosity as well as improving materials design capability in Material Science and Engineering.

Besides giving many multiple-choice questions as well as questions of short and long answer types marked in two categories following lower and higher order of Bloom's taxonomy, a list of references and suggested readings are given in the unit so that one can go through them for further advanced knowledge.

This section also highlights the examples of some interesting facts, applications of the subject matter for industrial applications on variety of aspects and/or our day-to-day real life, social and ethical issues (whichever applicable), and finally inquisitiveness associated with the topics of the unit.

RATIONALE

This unit helps students to get a primary idea about various nonferrous metals and their alloys. Several important alloys of aluminium, copper, magnesium, nickel, and titanium are described along with their binary equilibrium phase diagrams. Further, the properties and application of these alloys are also discussed along with bearing alloys. Important national and international standards have been discussed to further enhance the knowledge of the students.

The practical applications are related to the determining the suitable chemical composition and process parameter optimization during metal working. Finally, student will be capable to correlate the application of phase diagram in respective industries, and/or to acquire the knowledge of phase transformations in a metallic system.

PRE-REQUISITES

Crystal structure and Phase diagram

UNIT OUTCOMES

List of outcomes of this unit is as follows:

U3-O1: Describe binary phase diagrams of common non-ferrous system.

U3-O2: Acquire knowledge about various non-ferrous alloys and their applications.

U3-O3: Know about the various standards used for commercial alloys.

Unit-3 Outcomes	EXPECTED MAPPING WITH COURSE OUTCOMES (1- Weak Correlation; 2- Medium correlation; 3- Strong Correlation)				
	CO-1	CO-2	CO-3	CO-4	CO-5
U1-O1	3	1	3	1	1
U1-O2	3	1	3	1	2
U1-O3	3	1	3	1	1

INTRODUCTION

The bronze age and iron age are two important metallic periods in human civilization. Both ages are named after the artifacts (tools and weapons) that were made from these materials during that period. Bronze is a non-ferrous material which was discovered long back, even before iron. Now a days, non-ferrous metals and alloys have become of great importance in engineering applications. These are industrially very much relevant, and the practical applications of these materials are technologically important to improve materials design capability in material science and engineering.

3.1 ALUMINIUM AND ITS ALLOYS

Aluminium, a ductile and malleable metal with a density of 2.7 g/cc, is lighter than steel (7.8 g/cc) and copper (8.8 g/cc). In addition to having a low density, aluminium has excellent thermal and electrical conductivity. Aluminium forms a natural protective layer of aluminium oxide (Al_2O_3) over its surface due to its greater affinity for oxygen and thus offers superior corrosion resistance. The melting point of aluminium is 660°C. Although pure (unalloyed) aluminium has poor mechanical properties, alloying with other elements such as Si, Mg, Zn, Cu, Mn, Ti, etc. can enhance its mechanical abilities. The addition of alloying elements enhances the strength, weldability, workability, and corrosion resistance of the material. -

Copper has been generally considered as the principal alloying element in aluminium to provide a basis for precipitation (age) hardening and to decrease shrinkage and hot shortness. The aluminium-rich end of the aluminium-copper equilibrium phase diagram is shown in Fig. 3.1. The maximum solubility of copper in aluminium is 5.65 wt.% at 548°C. At room temperature, less than 0.25 wt.% Cu is soluble in aluminium. It is employed in amounts up to 4% in wrought alloys and up to 8% in castings.

Silicon is a very important alloying element for casting aluminium alloys. It improves casting qualities (fluidity and freedom from hot shortness), corrosion resistance, low thermal expansion, and high thermal conductivity. The aluminium-rich end of the aluminium-silicon equilibrium phase diagram is shown in Fig. 3.2. The maximum solubility of copper in aluminium is 1.65 wt.% at 577oC. At room temperature, negligible amount of silicon is soluble in aluminium. It is employed in amounts ranging from 1 to 14% as primary or secondary alloying elements.

3.1.1 Designations System in Aluminium Wrought Alloys

Aluminium and its alloys are divided into several classes or series depending upon the alloying element, purity, treatment etc. Aluminium wrought (mechanically worked) alloys are identified by a four-character code, with the first number denoting alloy group and the second indicating alloy modification. The last two digits identify the specific aluminium alloys or indicate the aluminium purity. Apart from the four digits, it also has some suffix which indicates the primary treatment and condition.

1xxx – The purity of the aluminium used to make these alloys is 99 per cent or higher.

2xxx – In this series, copper is the primary alloying element. Their strength can be enhanced by heat treatment. These alloys are tough and strong, although they are not as resistant to corrosion as other aluminium alloys.

3xxx – In this series, manganese is the primary alloying element. The most often used alloy in this category is 3003, which is both workable and somewhat strong.

4xxx – To make 4xxx series alloy, silicon is added to aluminium. This causes a reduction in the melting point of the metal without making it brittle.

5xxx – The primary alloying element in the 5xxx series is magnesium. Characteristics of these alloys include high strength, high weldability, and better corrosion resistance.

6xxx – The 6xxx series alloys are composed of silicon and magnesium. The elements combine to form magnesium silicide. These alloys have better corrosion resistance and moderate strength and are formable, weldable, and heat treatable.

7xxx – The major alloying element in this series is zinc. These alloys are heat-treatable and are very strong too.

8xxx – Other elements.

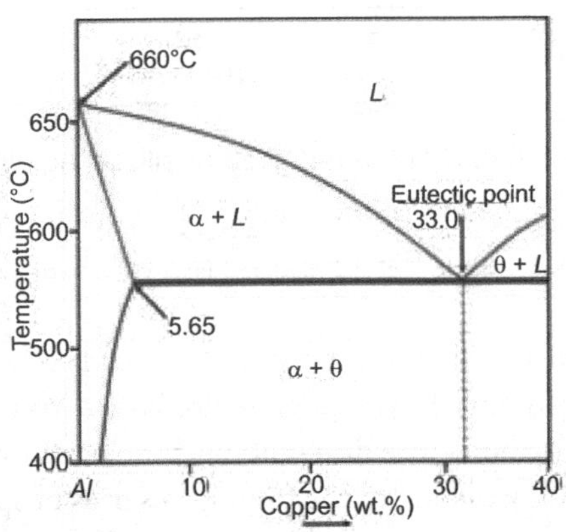

Fig. 3.1: Aluminium-rich end of the aluminium–copper equilibrium phase diagram

The mechanical properties of typical Aluminium alloys are given in Table 3.1. The suffix such as O, F, H and T are used to define the condition of aluminium alloys namely annealed, as fabricated, cold worked, and heat treated respectively. The number along with the T (heat treated) indicates the types of heat treatment operation as mentioned in Table 3.2. The – H is also followed by two-digit number. The first digit indicates the specific combination of thermomechanical treatment (1 for strain hardening only, 2 for strain hardening followed by partial annealing, and 3 for strain hardening followed by stabilization). The second digit (from 1 to 9) indicates the amount of cold working.

3.1.2 Special Aluminium Alloys

Duralumin: The most often used deformed aluminium alloy that can be strengthened by heat treatment is duralumin. "Duralumin" is a trademark, without a suffix, and does not indicate the alloy. Aluminium, copper, magnesium, manganese, silicon, and iron are at least six of the constituents of the alloy known as duralumin, although the main alloying components is copper. It typically has 4% Cu, and 0.5 to 1.5% Mg (and / or Mn) and small amounts of Fe and Si as well.

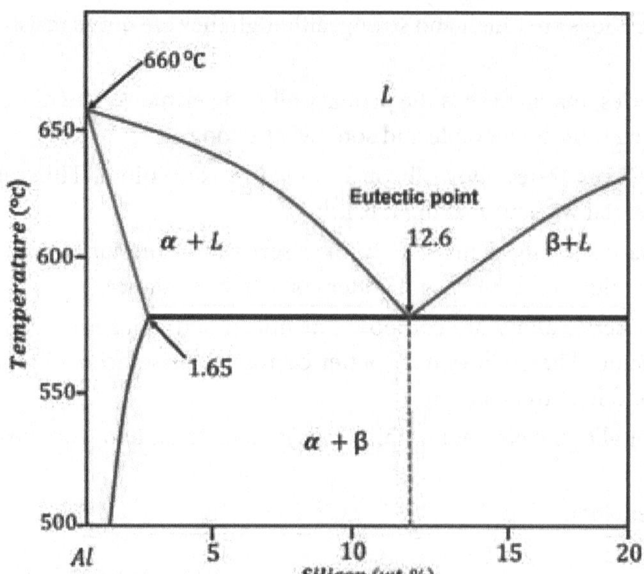

Fig. 3.2: Aluminium-rich end of the aluminium–silicon equilibrium phase diagram

The alloying element/s in the aluminium can create a variety of soluble compounds, including Mg_2Si, $CuAl_2$, and S-phase. It possesses strong mechanical and shock resistance capabilities after precipitation hardening treatment and good corrosion resistance as well.

Hiduminium: The term "Hi-Du-Minium" refers to *High Duty Aluminum* Alloys. This class of alloys were developed for aircraft use by Roll-Royce (RR) before World War II and hence they were described with the prefix RR. In order to manufacture different machine parts (such as pistons and cylinder heads of internal combustion engines) by casting or die forging, a series of specific aluminium alloys (Y-alloy/ RR-58) with additions of copper, magnesium, nickel, iron, and titanium are employed. These alloys are given the necessary characteristics by artificial ageing and hardening. The principal alloying elements are copper (Cu-4%), nickel (Ni-2%), magnesium (Mg-1.5%), and small quantities of iron and titanium. It has very high resistance to corrosion and has an excellent capacity to sustain strength at increased temperatures.

Hindalium: It is a common trade name of an alloy of aluminium, magnesium, manganese, silicon, and chromium. Hindalium is much stronger and costlier than aluminium. However, it is less costly than stainless steel. This alloy possesses good scratch resistance, strength and hardness, better corrosion resistance and can be easily cleaned. Therefore, hindalium is mainly used for manufacturing of anodised utensils.

Magnalium: It is an alloy of aluminium and magnesium. It has superior strength and is lighter than pure aluminium. This alloy possesses poor castability, better machinability and good weldability than aluminium. The molten metal is easily oxidised, and a lot of difficulties are encountered in the casting of these alloys. Hence, it is melted and cast in vacuum to tackle the foundry related issues associated with these alloys. Magnalium (with 3–6 wt.%Mg) is used for moderately stressed craft fittings, metal mirrors, jewellery, coins, automobile parts and scientific instruments. Magnalium with high amounts of magnesium (around 50%) are very brittle and more prone to corrosion than aluminium, which makes them unsuitable for structural applications. These alloys are used in fireworks applications, ferro alloys, and foundry flux. Magnalium nano powder is being reported as an additive to improve combustion, performance, and emission of biodiesel in direct injection diesel engine.

Choices of the alloys are governed by the intended application, cost and ease of manufacturing processes. In addition to the above, many other alloy compositions of aluminium and its alloys are also possible which are being used in various industries.

3.2. COPPER AND ITS ALLOYS

Since ancient times, copper and copper-based alloys have been used in a wide range of applications due to their desired combinations of physical properties. The melting point of copper is 1084°C.

Table: 3.1: Mechanical properties and applications of Aluminium alloys.

Alloys Name	Composition	Condition /Heat treatment	Tensile Strength (MPa)	Yield strength (MPa)	% elongation	Applications
AA1100	99%Al+0.12%Cu	Annealed	89	24	25	Sheet metal work
AA2024	4.4 Cu+1.5Mg	Annealed	220	97	12	Aircraft structure
		T6	442	345	5	
AA3003	1.2 Mn	Annealed	117	34	23	Pressure vessels

AA5052	2.5 Mg+0.25 Cr	Annealed	193	65	18	Automobile
		H14	262	179	4	
AA6061	1Mg+0.6 Si+0.27 Cu	Annealed	152	82	16	Structural application
		T6	290	241	10	
AA7075	5.6Zn +2.5Mg+1.6Cu	Annealed	276	145	10	Aircraft
		T6	504	428	8	

Table: 3.2: Shows the designation used in the heat treatment of aluminium alloys.

Heat Treated (T)	Operation
T-1	Partial solution and natural ageing
T-2	Annealed cast product only
T-3	Solution and cold work
T-4	Solution and natural ageing
T-5	Artificial ageing only
T-6	Solution and artificial ageing
T-7	Solution and stabilizing
T-8	Solution, cold work, and artificial ageing

The essential properties of copper are superior corrosion resistance, excellent electrical and thermal conductivity, and simplicity of fabrication. The most commercially available copper alloys can be classified into the following groups:

3.2.1 Brasses

Brasses are mostly made of alloys of copper and zinc. Alteration of the composition will result in the formation of desired strength, machinability, corrosion resistance, or a combination of these properties. The phase diagram of the copper-zinc system is shown in Fig. 3.3. The solubility of Zinc in copper increases from 32.5% at a temperature of 903 °C to 39% at 468 °C. While in most cases, the percentage elongation of the brass alloy decreases with increase in the solid solution element, but in some alloy combinations, increasing the zinc content enhances the ductility along with strength as well. The best combination of ductility and strength has been found to be at 70Cu-30Zn composition.

Based on the percent of zinc in copper, the brass designations have been further categorized into three groups:

(i) Alpha brasses: containing up to 36 % Zinc

   (a) Red brasses: containing 5 to 20% Zinc

   (b) Yellow brasses: containing 20 to 36 % Zinc

(ii) Alpha plus beta brasses: containing 54 to 62 % zinc

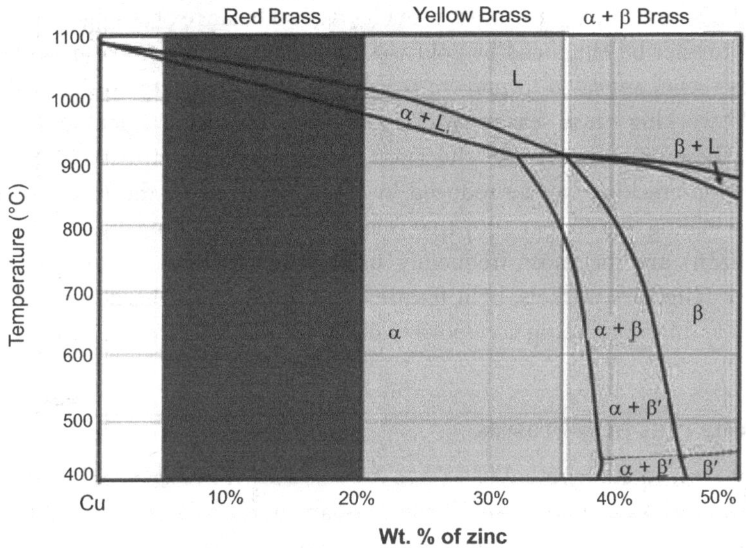

Fig. 3.3: Phase diagram of copper–zinc system.

3.2.1.1 Alpha brasses

Alpha brasses with up to 36% zinc provide good functioning characteristics and a relatively decent level of corrosion resistance. Depending on the quantity of copper present, alpha brasses occur in various colours, ranging from red for high-copper alloys to yellow for those with around 62 % copper. Yellow brasses and red brasses are two categories into which the alpha brasses can be subdivided.

(i) Red brasses

Red brasses have a zinc content of between 5% and 20%. They often withstand corrosion better than yellow brasses and resist dezincification (i.e. loss of zinc leads to formation of spongy and porous copper) and season cracking (i.e. a type of stress corrosion cracking resulting due to the combined effects of corrosion and internal stress). However, stress corrosion cracking (SCC) is a type of failure in which cracking occurs under combined action of corrosion and stress, either internal or external. Gilding metal (95Cu-5Zn), commercial bronze (90Cu-10Zn), red brass (85Cu-15Zn), and low brass i.e. 80Cu-20Zn are the most popular.

Gilding metal (95Cu-5Zn), which has more strength than copper, is used to make plaques, primers, emblems, fuse caps, coins, medals, and other items that are highly polished or gold-plated. Commercial bronze (90Cu-10Zn) is used for marine hardware, forgings, rivets, screws, and costume jewellery. It has good cold-working and hot-working qualities. For automatic screw machine work, screws and other components are made of leaded commercial (1.75% Pb). Electrical conduit, screw shells, sockets, hardware, condenser and heat exchanger tubes, plumbing pipe, lipstick cases, compacts, nameplates, tags, and radiator cores are all made of red brass (85Cu-15Zn). For decorative metalwork, medallions, thermostat bellows, musical instruments, flexible hoses, and other deep-drawn items, low brass (80Cu-20Zn) is utilized.

(ii) Yellow alpha -brasses

Zinc content of 20–36% in yellow α-brass provides a good balance of strength and ductility. Mechanical properties can further be enhanced by cold working. After prolonged cold working, these brasses are frequently stress-relief annealed to prevent season cracking. The large residual strains left in the brass because of cold working cause season cracking or stress-corrosion cracking. Particularly in ammonia environments, these pressures render the brass more vulnerable to intergranular corrosion. The risk of stress-corrosion cracking can be reduced by reducing stress in the recovery range (up to roughly 260°C) or by switching to a copper alloy that is less vulnerable. Yellow brass (65Cu-35Zn) and cartridge brass (70Cu-30Zn) are the most frequently used yellow α-brasses. Radiator cores of automotive, tanks, headlight reflectors; sockets, light fixtures, and screw shells; hardware such as eyelets, fasteners, grommets, and springs; plumbing accessories; and ammunition parts are just a few examples of typical uses.

3.2.1.2 Alpha plus beta brasses

These brasses have a copper content of 54% to 62%, as shown in Fig. 3.3 which demonstrates that these alloys will have two phases namely α+β. These alloys are more challenging to cold work than the α brasses because the α+β phase is harder and more brittle at room temperature. Since most of these alloys can be heated into the single-phase β region, where the β phase becomes very plastic at high temperatures, they offer good hot-working characteristics.

Muntz metal (60Cu-40Zn), which has high strength and exceptional hot-working characteristics, is the most popular kind of α+β brass. The β region's rapid cooling may suppress the majority of the α phase's precipitation. Ship sheathing, condenser heads, perforated metal, and architectural work: all employ Muntz metal in sheet form.

3.2.2 Bronzes

The term bronze, in general refers to the copper and tin alloy system. However, it is now also used for any copper alloy (excluding copper zinc alloys) containing up to 1.2 percent of the major alloying element. Commercial bronzes are mostly copper alloys containing tin, aluminium, silicon, or beryllium. They may also include phosphorus, lead, zinc, or nickel.

3.2.2.1 Tin bronzes

Tin bronze contains 1 to 11% tin in copper. However, a little amount of phosphorus is also required (up to 0.01-0.5%) which acts as a deoxidizer in casting. Figure 3.4 depicts the copper-rich section of the copper-tin alloy system. At 798°C, a peritectic reaction produces the β phase.

The beta phase initiates a eutectoid reaction at 586°C to generate the eutectoid mixture (α+γ). Gamma (γ) similarly undergoes a eutectoid transition to (α+δ) at 520°C. The diagram also illustrates the delta (δ) phase decomposition. This is accomplished by a 350°C eutectoid reaction that produces (α + ε). This process is so slow that the epsilon (ε) phase does not occur in commercial alloys.

Below 520°C, the slope of the solvus line demonstrates a significant drop in tin solubility in the Cu-Sn system (see Fig. 3.4). Different types of precipitates will form as a function of temperature due to this difference in solubility as indicated by the solvus line below 520°C.

As a result, slow-cooled cast tin bronzes having less than 7% tin often exhibit only a single phase, the solid solution. Most castings having more than 7% tin have some of the 6 phases. The structure of a quickly cooled cast 10% phosphor bronze reveals tiny phase particles in a thin dendritic matrix.

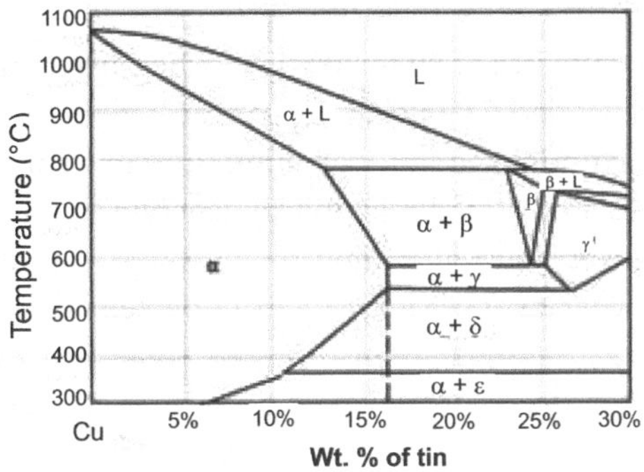

Fig. 3.4: Phase diagram of copper–tin system.

Phosphor bronzes, alloys of Cu, Sn and P are widely used due to their excellent strength, toughness, strong corrosion resistance, low friction coefficient and resistance to freedom from season cracking. They're commonly utilised in membranes, bellows, lock washers, cotter pins, bushings, clutch discs, and springs.

Sometimes zinc is used to replace some of the tin. As a result, casting characteristics and toughness improve while wear resistance remains unchanged. Tin bronze is frequently alloyed with lead to increase machinability and wear resistance. Lead content in high-lead tin bronze can reach up to 25%. Leaded alloys are employed in bushings and bearings that are subjected to moderate or low stresses.

3.2.2.2 Silicon bronzes

Silicon bronze is copper rich alloy (about 95% Cu) with balance mostly comprised of silicon and other elements (Sn, Zn etc.) in much smaller quantities. Figure 3.5 depicts the copper-rich region of the copper-silicon alloy system. Silicon solubility in the phase is 5.3 percent at 852°C and declines with temperature. Since the eutectoid reaction at 554°C is relatively slow, commercial silicon bronzes, which typically contain less than 5% silicon, are single-phase alloys. The strongest of the work-hardenable copper alloys are silicon bronzes. They have mechanical qualities like mild steel and corrosion resistance similar to copper.

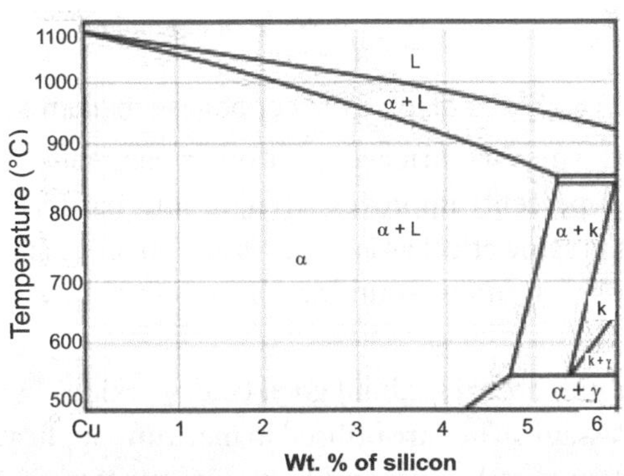

Fig. 3.5: Phase diagram of copper–silicon system.

They are utilised in the building of tanks, pressure vessels, maritime structures, and hydraulic pressure lines.

3.2.2.3 Aluminium bronzes

Aluminium bronze typically comprises of 80 to 90% Cu while the remaining elements are comprised of mainly Al, and the balance are Ni and Fe. Fig.3.6 depicts the copper-rich area of the copper-aluminium alloy system. At 565°C, the maximum solubility of aluminium in a solid solution is around 9.5 percent. At 565°C, the phase undergoes a eutectoid reaction to generate the $(\alpha+\gamma_2)$ mixture. Most commercial aluminium bronzes have between 4 and 11% aluminium. Alloys with up to 7.5 percent aluminium are normally single-phase alloys, whereas those with 7.5 to 11 percent aluminium are two-phase alloys.

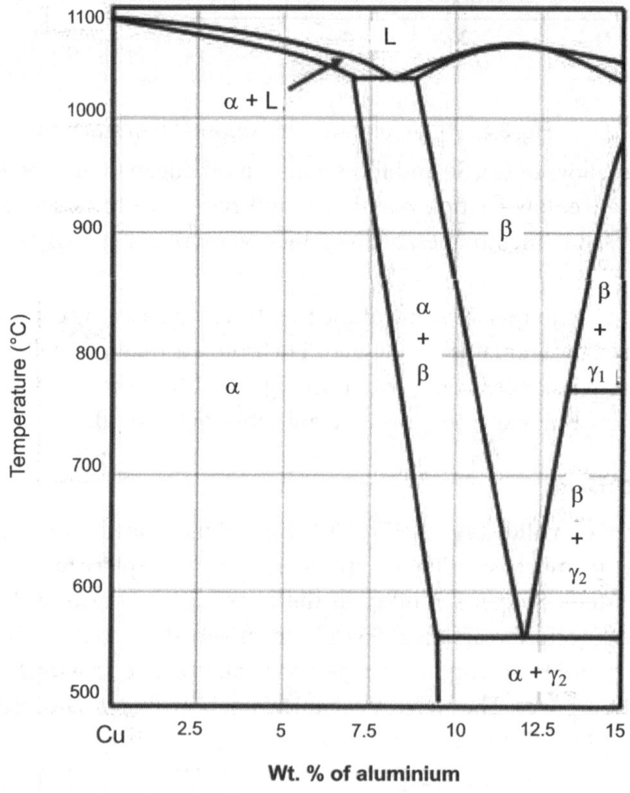

Fig. 3.6: Phase diagram of copper–aluminium system.

Aluminium bronzes are typically enriched with other elements such as iron, nickel, manganese, and silicon. Iron (0.5 to 5.0 percent) improves strength and hardness while also refines the grains; nickel (up to 5 percent) has the same effect as iron but is less effective; silicon (up to 2 percent) improves machinability; and manganese promotes soundness in castings by combining with gases while also improving strength.

The single-phase aluminium bronzes exhibit good cold-working characteristics and strength, as well as resistance to air and water assault. They are utilised in maritime applications for condenser tubes, cold-work forms, corrosion-resistant vessels, nuts and bolts, and protective sheathing. The α+β aluminium bronzes are attractive because they can be heat treated to obtain desired structures similar to those seen in steel. A lamellar structure resembling pearlite is created when the furnace cooling is performed from temperatures exceeding the eutectoid temperature. When the two-phase alloy is quenched from 815-871°C, a needle like structure resembling martensite is produced. To enhance strength and hardness, the quenched alloys are tempered between 371 and 593°C. Heat-treated aluminium bronzes are utilised in gears, propeller hubs, blades, pump components, bearings, bushings, non-sparking tools, and drawing and shaping dies.

3.2.2.4 Beryllium bronzes

Beryllium bronze is copper rich Be-Cu alloy with Be contents generally varying between 0.5 – 3.0%. Figure 3.7 depicts the copper-rich region of the copper-beryllium alloy system. Beryllium's solubility in a solid solution fall from 2.1 % at 865°C to less than 0.25 percent at ambient temperature. This alteration in solubility is always suggestive of the possibility of age hardening.

The optimum mechanical properties are obtained in alloys containing around 2% beryllium. For this alloy, a typical heat-treating cycle would be: solution at 788°C, water-quench, cold-work, and finally ageing at 315°C. Beryllium bronzes are utilised for springs, bolts, screws, firing pins, dies, and other applications because of their good formability, high yield strength, and great wear and corrosion resistance. Cu-Be alloys are suitable for explosion-proof tools as they do not produce sparks when impacted.

3.2.3 Cupro-Nickels

These are copper-nickel alloys with up to 30% nickel content. Because the copper-nickel binary phase diagram (Fig. 2.8) demonstrates complete solubility, all cupro-nickels are single-phase alloys. They cannot be heated, hence the only method to modify their properties is by cold working.

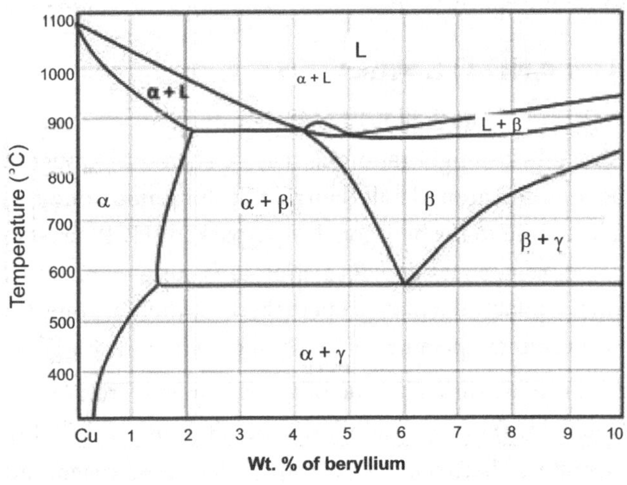

Fig. 3.7: Phase diagram of copper–beryllium system.

Cupronickel alloys have a strong resistance to corrosion fatigue in addition to being very resistant to the corrosive and erosive effects of rapidly moving sea water. They are widely used for condenser, distiller, evaporator, and heat exchanger tubes aboard naval ships and coastal power facilities.

3.2.4 Nickel Silvers

Nickel silver alloys are copper, nickel, and zinc ternary alloys. They are so called due to their silvery colour in appearance and not due to the presence of real silver (Ag) in the alloys. Commercial alloys are made up of 5-40% zinc, 5-30% nickel, and 50-70% copper. At room temperature, single-phase nickel silver alloys having more than 60% copper are ductile and easy to work with, but they only have fair hot-working qualities. The inclusion of nickel imparts an attractive silver-blue white colour to the copper-zinc alloy, as well as strong corrosion resistance to water, air, and food chemicals. These alloys yield excellent base metals for chromium, nickel, or silver plating. They are used to make radio dials, nameplates, crockery, flatware, zippers, costume jewellery, and rivets and screws. Nickel silver alloys with 50-60% copper are $\alpha+\beta$ alloys with two phases. The phase diagram for Cu-Ni-Zn system has not been given as ternary phase diagram is beyond the scope of this textbook. Nickel silvers have a high modulus of elasticity and, like $\alpha+\beta$ brasses, may be hot worked readily. Nickel silvers have a lower stress corrosion resistance than binary copper-zinc alloys with the same quantity of zinc. Typical uses for $\alpha+\beta$ nickel silvers include telephone equipment springs and connections, resistance wire, hardware, and surgical and dental equipment.

3.2.5 Designation of Copper Alloys

The United States' Copper Development Association (CDA) has categorised the recognition system for copper alloys. Cast copper alloys are labelled from C1xxxx to C7xxxx, while wrought copper alloys are named from C8xxxx to C9xxxx. According to the CDA designation system, the primary alloying element in copper alloys is represented by the first number in the table 3.3 and typical mechanical properties of copper alloys are tabulated in Table 3.4.

3.3 MAGNESIUM AND ITS ALLOYS

The main benefits of magnesium include its lowest density of all structural metals (1.7 g/cc), ease in machining, and high strength-to-weight ratio that may be achieved through its alloys. Based on equal volumes, weight of magnesium is around half than that of aluminium, one-fourth of steel, and one-fifth of copper and nickel alloys. Due to the hexagonal close-packed (HCP) crystal structure of magnesium, slide along the basal planes causes plastic deformation to occur at ambient temperature. Compared to FCC metals, magnesium has poorer ductility property because there are only three slip systems accessible for plastic deformation at room temperature. The melting point of magnesium is 650°C.

Magnesium and its alloys are difficult to deform at room temperature; in fact, only minor amounts of cold work may be applied without annealing. As a result, most of the fabrication is done by casting or hot working at temperatures between 200 to 350 °C. However, magnesium is difficult to cast due to its high affinity with oxygen and hence either flux or protective environment is required during its melting and casting. Pure magnesium has better corrosion resistance, but magnesium alloys show poor corrosion resistance due to the formation of micro galvanic cells. More than one alloying element is added to magnesium to get better mechanical properties and superior corrosion and wear resistance.

Table 3.3: The CDA designation system of copper alloys.

C1	Pure copper
C2	Brasses
C3	Lead brasses
C4	Tin brasses
C5	Phosphor bronze
C6	Aluminium and silicon bronze
C7	Nickel silvers
C8	Cast copper alloys
C9	Cast copper tin alloys

Table 3.4: Typical mechanical properties of copper alloys

Alloy No.	Chemical composition	Condition	Tensile strength (MPa)	Elongation %
C10100	99.99Cu	Annealed	220	45
		Cold worked	345	6
C26000	70Cu+30Zn	Annealed	325	62
		Cold worked	525	8
C28000	60Cu+40Zn	Annealed	370	45
		Cold worked	485	10
C17000	99.5Cu + 1.7Be + 0.2 Co	Solution treated	410	60
		Cold worked	1240	4
C61400	91Cu+7Al+2Fe	Annealed	550	40
		Cold worked	615	32
C71500	70Cu+30Ni	Annealed	380	36
		Cold worked	580	3
C80500	99.75Cu	As cast	172	40
C90300	88Cu+8Sn+4Zn	As cast	310	30

3.3.1 Designation of Magnesium Alloys

Magnesium alloys are designated by 3-5 alphanumeric codes. In that code, the first two letters provide information about the largest proportion of alloying elements on a decreasing basis. Designation of the first two letters are follows: A (Aluminium), B (Bismuth), C (Copper), D (Cadmium), E (Rare earth), F (Iron), G (Magnesium), H (Thorium), K (Zirconium), L (Beryllium), M (Manganese), N (Nickel), P (Lead), Q (Silver), R (Chromium), S (Silicon), T (Tin), Y (Antimony), and Z (Zinc). The letters are followed by a two-digit figure that, to the closest percent, represents the quantities of each of the two alloying elements. The last letter after two-digit number indicates the stage of development of the alloy i.e. A, B, C, D, or E. These letters also indicate the degree of purity. Additionally, temper specifications are employed for alloys of magnesium and aluminium. For example: AZ31B-F is a magnesium alloy, having aluminium and zinc as the major alloying element with 3% and 1% respectively. The designation of F is to represent 'as fabricated' state.

3.3.2 Magnesium Alloys

Magnesium based alloys are regarded as the next generation structural materials due to their low density and high strength-to-weight ratio, which provides considerable weight saving. However, the mechanical property of pure magnesium is very small. Further, it is difficult to deform magnesium due to the limited number of slip systems present in this material. Various alloying elements are used to tailor the mechanical, corrosion, wear, and other properties of magnesium. Fig. 3.8 shows typical applications of various magnesium alloys. The typical mechanical properties of magnesium alloys are tabulated in table 3.5.

As a result, they have great potential especially as light weight aerospace materials, transportation structures, casing of computers, communications, and consumer electronic devices.

3.3.2.1 Magnesium-aluminum based alloys

Casting alloys made of magnesium, aluminium, and manganese (AM) and aluminium, and zinc (AZ) are included in this category. With an excellent balance of tensile strength, yield strength, and elongation, the AM100A alloy is well-liked for pressure-tight sand and permanent mould castings. For normal temperature applications, the sand-casting alloys AZ63A and AZ92A are utilised. They may perform excellently at temperatures as high as 176°C if the operating stresses are not excessive. When maximal ductility or toughness is needed together with a moderately high yield strength, AZ63A is employed. When high yield strength and good pressure-tightness are required, AZ92A is employed.

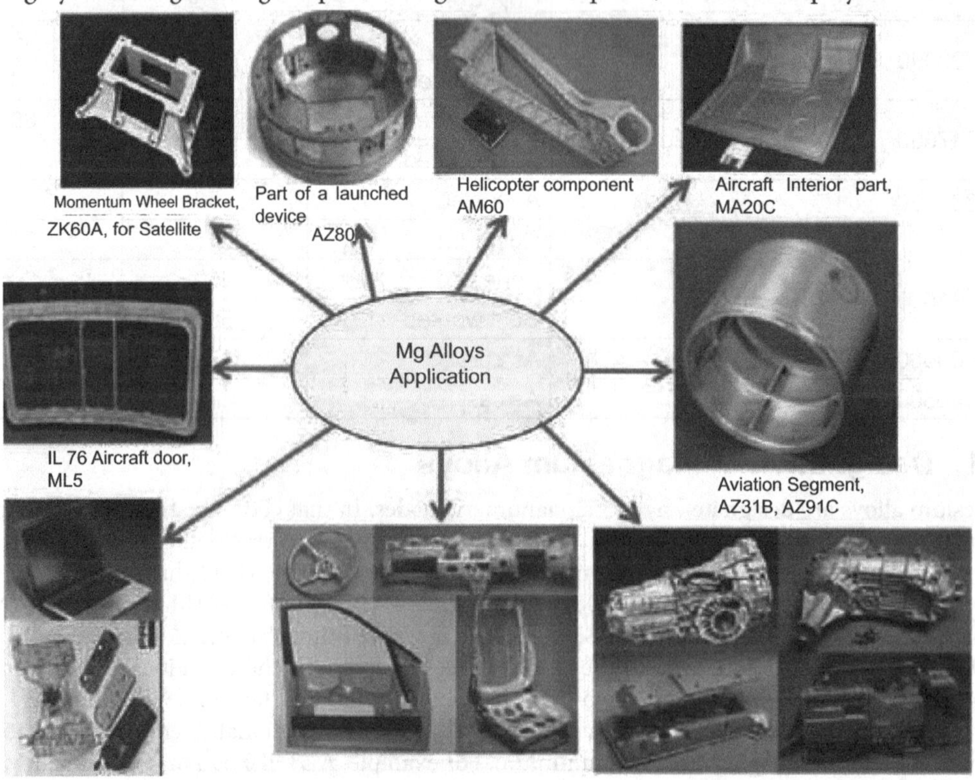

Fig. 3.8: Typical applications of magnesium alloys

As all-purpose extrusion alloys with high strength and formability, AZ31 alloys are commonly employed. They are strain hardened to acquire their qualities because they only contain 3% aluminium and cannot be heat treated.

AZ31 with low calcium content, known as PE alloy, is widely used for magnesium photo-engraving sheets in the printing industry. AZ61A has excellent strength and ductility and is used mainly as an extrusion and forging alloy. AZ80A alloy, containing 8.5% aluminium, is a heat-treatable alloy used for extruded products and press forgings.

3.3.2.2 Magnesium-zinc based alloys

Essentially, these are casting alloys made of magnesium, zinc and zirconium (ZK), and magnesium, zinc, and thorium (ZH). Zirconium additions to magnesium alloys, up to around 0.7 percent, have a grain-refining effect. The columnar cast structure is eliminated by zirconium, thereby improving the mechanical characteristics. The tensile strength and ductility of the magnesium casting alloys ZK51A and ZK61A are combined to the utmost degree. This is caused by zirconium's fine-grain effect and the significant age-hardening impact of the magnesium-zinc binary system. To create sound castings free of microporosity and hot fractures, the high zinc concentration unfortunately necessitates extremely rigorous foundry management. Both the inclusion of a rare-earth metal (ZE41A) and the addition of thorium (ZH62A) contribute to the mitigation of the mentioned issues while decreasing the mechanical characteristics. The strongest extrusion alloy for solid and hollow structures is the wrought alloy ZK60A. It possesses decent hardness and can be heat-treated to further enhance its qualities. Zirconium is switched out for a rare-earth metal (ZE10A), creating a stronger and more affordable alloy.

3.3.2.3 Magnesium-rare earth based alloys

The magnesium-rare earth-zirconium (EK) and magnesium-rare earth-zinc (EZ) casting alloys are included in this category. The atomic numbers of the rare-earth elements are 58–71. A high recrystallization temperature and precipitates at the grain boundaries that are stable at high temperatures to reduce creep are used to improve elevated-temperature characteristics. Both criteria are met by the addition of rare-earth elements to magnesium, making these alloys useable up to 260°C.

Table 3.5: Typical mechanical properties of the magnesium alloys

Alloy Name	Composition (balanced by Mg)	Condition/ Treatment	Tensile strength (MPa)	Yield strength (MPa)	% of Elongation
AZ31B	3Al + 1 Zn	Annealed	228		11
HM21A	2Th+ 0.8Mn	T8	228	138	6
AZ63A	6Zn + 3Al	cast	179	76	4
AZ91C	9Al+ 1Zn	T6	275	131	5
ZK60	6Zn + 0.5Zr	T5	310	235	5

The mechanical properties of the casting alloys EK30A, EK41A, and EZ33A are comparable, however due to the addition of zinc, EZ33A exhibits less significant corrosion resistance. No wrought-magnesium alloys include the rare-earth elements as the main component.

3.3.2.4 Magnesium-thorium based alloys

The magnesium thorium-zirconium (HK) and magnesium thorium-zinc (HZ) casting alloys are included in this category. Like the rare-earth elements, thorium significantly enhances magnesium's high-temperature capabilities. HK31A and HZ32A alloys are employed for applications where working temperature falls between 176oC and 371°C. HZ32A is chosen when long-term, lower-stress qualities are crucial, while HK31A is best for usage involving brief periods of rising temperature and high stresses. As a wrought sheet alloy for high-temperature applications, HK31A is also employed. The extrusion alloy HM31XA and the sheet alloy HM21A, both magnesium-thorium-manganese alloys, are other wrought alloys in this category. These are the wrought-magnesium alloys with the best high-temperature characteristics.

3.4. NICKEL AND ITS ALLOYS

Nickel is well-known for its corrosion and oxidation resistance. The melting point of nickel is 1455oC. It has a white appearance and excellent workability and mechanical properties. Stainless and nickel-alloy steels require around 60% of the nickel production. Most of the remaining nickel is utilised in high-nickel alloys and electroplating. Due to its superior corrosion resistance and hardness, nickel also serves as an effective corrosion and wear resistant coating material. A nickel, D nickel, E nickel, permanickel, and duranickel are the most common commercial grades of nickel. The typical mechanical properties of nickel alloys are tabulated in table 3.6.

Copper, iron, chromium, silicon, molybdenum, manganese, and aluminium are the elements most often alloyed with nickel. For example, in Nickel-Copper alloys, copper is entirely soluble in nickel and is used to enhance formability, reduce costs, and retain nickel's corrosion resistance as well.

3.4.1 Nickel and Copper Alloys

Monel is the most important of the nickel-copper alloys, containing approximately two-thirds nickel and one-third copper. Monel has high corrosion resistance to acids, alkalis, brines, waters, food products, and the atmosphere. It has mechanical properties higher than those of the brasses and bronzes, but lower than those of alloy steels. It also has good toughness and fatigue strength and finds considerable use in elevated-temperature applications. It does not oxidize at a destructive rate below approximately 725°C in sulphur-free atmospheres and for some applications may be used at temperatures as high as up to 1250°C. Monel has widespread use in the chemical, pharmaceutical, marine, power, electrical, laundry, textile, and paper-equipment fields.

3.4.2 Nickel-Silicon-Copper based Alloys

The commercial name for nickel-silicon-copper alloys is Hastelloy D. This alloy is formed by the casting technique and has a chemical composition of 10% Silicon and 3% Copper. It has extremely good mechanical qualities, such as being very hard and tough, as well as having exceptional corrosion resistance in a sulfuric acid environment at higher temperatures. It is used in the chemical industry for evaporators, reaction vessels, pipelines, and fittings.

Table 3.6: Mechanical properties of Ni based alloys

Alloys	Composition	Condition/ treatment	Tensile strength (MPa)	Yield strength (MPa)	% elongation
Monel	66.15 Ni+31.3Cu	Annealed	517	241	40
		Cold rolled	758	690	5
Inconel	72.85Ni+15.15Cr+6.8 Fe	Annealed	692	344	50
Hastelloy A	Ni+22Mo+22Fe	As cast	503	303	10
Hastelloy D	Ni+10Si+3Cu	Sand cast	813	813	2

3.4.3 Nickel-Chromilum-Iron based Alloys

A wide range of the nickel-chromium binary system and nickel-chromium-iron ternary system alloys are employed to be used as electrical-resistance materials. The 80Ni-20Cr binary alloys (also known as *Nichrome or Chromel A*) are utilized in domestic appliances and industrial furnace materials as heating elements. The 60Ni-16Cr-24Fe ternary alloys (Chromel C, Nichrome, etc.) are used in immersion rods, hair dryers and toasters as electrical heating elements. Because of their high temperature (982°C) oxidation resistance and superior thermal fatigue properties, these alloys are also effective in heat treatment equipment and components for carburizing or cyaniding.

Inconel alloys (76Ni-16Cr-8Fe) are corrosion-resistant materials that are appropriate for usage in high-pressure, high-heat environments. When Inconel is heated to high temperatures, it forms a protective layer that protects the surface from further damage. Inconel retains high strength across a wide temperature range, making it a desirable material for high-temperature applications. Precipitation hardening or solid solution strengthening is the major strengthening mechanism of Inconel to offer high strength at increasing temperatures. The Inconel-625, Inconel-718, and Inconel-600 alloys are typically used in high temperature applications.

Inconel, with a nominal composition of 76Ni-16Cr-8Fe, combines the inherent corrosion resistance, strength, and toughness of nickel with the extra resistance to high-temperature oxidation of chromium. The initial applications for Inconel were in food-processing equipment such as heaters, coolers, regenerators, pasteurizers, and holding tanks for pasteurizing milk. It is also utilised in aircraft and heat treatment equipment due to its extraordinary ability to resist a broad range of temperatures without becoming brittle.

3.4.4 Nickel-Molybdenum- Iron based Alloys

Hastelloy M (57Ni-20Mo-20Fe) and *Hastelloy B (62Ni-28Mo-5Fe)* are the two best-known alloys in this group. Because austenitic alloys are not ideal for age hardening, the mechanical characteristics of these alloys can be improved by cold working. Because of their strong corrosion resistance in chemical environment, these alloys are particularly suitable for use in chemical industries. They are commonly employed in the chemical industry for handling, transporting, and storing acids and other corrosive compounds.

3.4.5 Nickel-Chromium-Molybdenum-Iron-based Alloys

The remaining Hastelloy alloys (54Ni-17Mo-15Cr-5Fe-4W) fall within this category, with Hastelloy being the most well-known. These alloys are distinguished by their great corrosion resistance to oxidising acids such as nitric, chromic, and sulfuric acids. They have strong high-temperature characteristics and are resistant to oxidising and reducing atmospheres up to 1725°C. They are employed in the chemical industry to serve in an environment containing strong oxidising acids, for pump and valve components, spray nozzles, and other related purposes.

3.5 TITANIUM AND ITS ALLOYS

Titanium has been employed for a variety of purposes since its discovery, in new and interesting industries as technology advances, including the aircraft and ship industries. It was discovered in 1791. The melting point of titanium is 1668oC. Titanium with 99.9% purity is commercially manufactured. It has trace amounts of contaminants such as carbon, iron, and oxygen. Unalloyed titanium is classified into four grades. The grades are arranged according to the relative corrosion resistance, ductility, and strength requirements for a given application. Grade 1 titanium provides the best corrosion resistance, ductility, and strength. Grade 4 offers the greatest strength and formability. Oxygen, carbon, and nitrogen impurities strengthen titanium by producing a solid solution. The density, melting point, elastic modulus, shear modulus and tensile strength of pure titanium are 4.5 g/cc, 1668oC, 110 GPa, 44 GPa, and 760-1100 MPa respectively.

Alloying elements are commonly referred to as alpha or beta stabilisers. The relative proportions of alpha and beta stabilisers in an alloy, as well as the heat treatment, determine whether its microstructure is mostly single-phase alpha, a combination of alpha and beta, or single-phase beta across its practical temperature range. They have significant influence on the properties of the product. The alpha to beta transition temperature is affected by the addition of these alloying elements to titanium. An alpha stabiliser raises the alpha to beta transition temperature when solute is added; a beta stabiliser reduces the transformation temperature. The alpha stabilizer is aluminium, whereas Cr, Mo, V, Mn, and Fe are used as beta stabilizers.

3.5.1 Alpha Alloys

Most of the alpha alloys contain some beta-stabilizing alloying elements. The compositions of these alloys are balanced by high aluminium content so that the alloys are essentially one-phase alpha. The key properties of alpha alloys are weldability and high-temperature strength retention. The first is due to the one-phase microstructure, while the second is due to the presence of aluminium. Alloying elements in solution strengthen alpha-phase alloys, and aluminium is the most effective strengthening agent for alpha alloy. Its impact is especially essential since it lasts even at high temperatures.

Hot working of alpha alloys containing more than about 6 percent aluminium is difficult. Hot workability of high-aluminium alpha is improved by additions of beta-stabilizing alloying elements in amounts small enough so that the beta phase is present in small quantities in the annealed microstructure.

3.5.2 Alpha-Beta Alloys

The beta phase occurs when there are enough beta stabiliser elements present. The addition of beta alloying in solution improves the properties of alpha beta alloys over the alpha phase. When the alpha phase of alpha-beta alloys is enhanced by aluminium, the alpha beta alloy becomes even stronger, particularly at high temperatures. The mechanical properties of typical alpha-beta alloy i.e. Ti-6V-4Al is tabulated in Table 3.7. Heat treatment can improve the mechanical characteristics of alpha beta alloys even further. Essentially, this is performed by quenching from an alpha-beta field temperature and then ageing at a considerably increased temperature. Unlike the traditional age-hardening technique, no homogenous beta solid solution is generated in the first stage. If an all-beta structure were formed, the beta grain size would be excessively large, and alpha would form primarily at the beta grain boundaries.

These two features lower the aged alloy's ductility. The transition of the elevated-temperature beta phase that would occur with slow cooling is prevented by quenching. Aging at high temperatures results in the precipitation of tiny (fine) alpha particles in volumes that were formerly beta grains. This type of very fine structure is stronger than the coarse annealed alpha-beta structure. In rare situations, quenched titanium alloy structures may be an unstable type of alpha known as titanium martensite.

When compared to the qualities of annealed material, heat treatment can enhance the strength of alpha-beta alloys by roughly 35%. The Ti-6AI-4V alloy is often used in aviation gas turbine compressor blades and discs, forged airframe fittings, and sheet metal airframe components.

3.5.3 Beta Alloys

Unlike the alpha alloys, beta alloys (Ti- 3AI-13V-11Cr) can be strengthened by heat treatment. This alloy is weldable in both the annealed and heat-treated conditions. Aging at elevated temperature after solution treatment results in the precipitation of fine particles of alpha and TiCr2 compound.

Ultimate strengths up to 1241 MPa with 6 percent elongation are possible after heat treatment. This is an increase over the annealed strength of at least 50 percent. Beta alloys have been used for high strength fasteners and for aerospace components requiring high strength at moderate temperatures.

Table 3.7: Typical mechanical properties of different Ti based alloys.

Alloys	Compositions	Condition	Tensile strength (MPa)	Yield strength (MPa)	% elongation
Pure Ti	Ti-99	Annealed	545	434	27
Alpha Ti	Ti-5Al-2.5Sn	Annealed	862	827	18
	Ti-6Al-4Zr-1V	Annealed	985	951	17
	Ti-8Al-1Mo-1V	Annealed	1013	930	16
Alpha-Beta Ti	Ti-8Mn	Annealed	951	862	15
	Ti-4Al-4Mn	Heat treated	1116	965	9
	Ti-6Al-4V	Annealed	930	827	111
	Ti-6Al-4V	Heat treated	1172	1034	7
Beta Ti	Ti-3Al-13V-11Cr	Heat treated	1241	1206	6

3.6 BEARING ALLOYS

Bearings are the most crucial components of any moving machine, which has the load bearing capability at high rotational or linear motion. It reduces friction. The bearing material should possess high fatigue strength and resistance to erosion and corrosion. Bearing materials are generally made up of lead-tin alloys (Babbit), and aluminium based alloys. The melting points of lead and tin are 327oC and 232oC respectively.

3.6.1 Babbit or White Metal Alloy

Babbit (or Babbitt) refers to the class of alloys commercially designated for the bearing application since the 19th century. Babbit alloy, also termed as white metal alloy are basically lead and tin based alloys with minor addition of copper or antimony. The lead based babbit can be categorized into two groups. One group corresponds to the lead-tin-antimony alloy with addition of arsenic while the other group include lead and tin with minor amounts of calcium, magnesium, barium, and sodium. The lead based babbit has high absorption capability and shows tolerance to misalignment. These alloys are mostly used *for automotive connecting rods, main and* camshaft bearings, diesel-engine bearings, and *electric motor* bearings. However, due to several environmental and safety concerns, the lead babbit finds restricted application. An alternative to the lead babbit is the tin babbit, which comprises of more than 80%Sn. Small amount of lead, antimony and copper are also present along with tin. The alloy consists of rod shaped CuSn and cubes of SnSb compounds present in a ductile tin-rich ternary eutectic. The SnSb cubes have extremely high hardness and contribute to the excellent wear and corrosion resistance of the alloy. The tin babbit are used for the high speed and heavy load bearing applications. Table 3.8 lists the composition and mechanical properties of the lead and tin babbits. Tin babbits have better corrosion and wear resistance than lead-based bearing alloys. The bureau of Indian standards (BIS) published IS:25-1979 covers the specification for ten grades of antifriction bearing alloys (Pb-based and Sn-based alloys).

3.6.2 Aluminium based Bearing Alloys

Aluminium is a lightweight metal and has high specific strength along with several combinations of bearing properties like good fatigue strength, wear resistance and anti-friction properties like conformability. Al based bearings are employed as connecting rods and main bearings in internal combustion engines. Aircraft landing gear assemblies, power shovels, and track rollers utilize the Al bearings to withstand high shock loads. Among the several Al alloys, Al-Si and Al-Sn-Si alloy have emerged as promising bearing alloys. The BIS published IS:6754-1972 and IS:6751-1972 which cover the requirements of aluminium based bearing alloys.

Aluminium-Silicon bearing: The Al-Si alloys possess high strength and high thermal conductivity at elevated temperatures. It comprises of a ductile Al matrix and brittle Si particles embedded in the matrix. Si is a hard phase in Al-Si bearing alloys and provides greater hardness and improves wear and seizure resistance. The alloy also exhibits good corrosion resistance. The poor resistance to seizure (sudden failure as a result the bearing cannot support motion and the system totally shuts down) makes Al-Si alloys vulnerable under poor lubricating conditions during starting or warming up of the engines. Sn is added to the Al-Si alloys as a solid lubricant to minimize the effect of seizure. The Sn content in the alloy is restricted between 10-20 wt% and Si content lies in the range of 2-11 wt%.

3.6.3 Bearing Bronze

Bronze (Cu-Sn) alloys show low-sliding friction, better wear, and corrosion resistance. There are four main types of bronze alloys namely aluminum bronze, manganese bronze, phosphorous bronze, and silicon bronze. All bronze alloys possess good machinability, wear and corrosion resistance, and have yield strengths ranging between 138–414 MPa.

Table 3.8: Chemical composition and mechanical properties of bearing materials

Alloy	Nominal compositions (wt%)				Mechanical properties		Applications
	Pb	Sn	Sb	Cu	Tensile strength (MPa)	Hardness (BHN)	
Lead babbits:							
SAE 13	85	5	10	-	69	19	Light loads: car journal bearings
SAE 14	75	10	15	-	72	22	Moderate loads: pumps, blowers
SAE 15	83	1	15	-	71	20	High loads: diesel engine bearings
Tin babbit	-	91	4.5	4.5	64	17	Automotive applications

The BIS published IS:3980-1982 which covers two types of porous metal powder oil-impregnated bearings namely Grade 1–Copper base (Class A: Cu-Sn; & Class B: Cu-Fe-Sn) and Grade 2–Iron base (Class A: Pure Fe; & Class B: Fe-Cu).

UNIT SUMMARY

1. The present chapter provides an insight about several important non-ferrous metals and alloys, corresponding phase diagrams, designation, heat treatments, and applications.
2. The aluminium alloys are designated based on the alloying elements and as per the heat treatment cycle, they find suitable applications. Duralumin, Hindalium *and* Magnalium are commercially famous aluminium alloys.
3. The commercially available copper can be classified mainly as brass, bronze, cupronickel, and other Cu based alloys based on the alloying elements. The phase diagram of all these alloys have been extensively discussed to suggest the importance of selecting a particular composition.
4. Magnesium alloys are also designated based on the alloying elements. The significance of Mg-Al, Mg-Zn, and Mg-Rare-Earth alloys has been provided based on the applications.
5. The Ni based alloys comprising of Cr and Fe is an important engineering material for high temperature applications. The 80Ni-20Cr binary alloys (also known as *Nichrome or Chromel A*) are utilized in domestic appliances and industrial furnace materials as heating elements. Similarly, *Inconel alloys (76Ni-16Cr-8Fe)* are corrosion-resistant materials that are appropriate for usage in high-pressure, high-heat environment.

6. Titanium alloys exhibit high temperature mechanical properties. The different alloys of titanium such as alpha, alpha-beta and beta alloys, their strengthening mechanisms and mechanical properties have been thoroughly discussed.
7. Different bearing alloys and relevant standards of BIS have also been discussed briefly.

EXERCISES

Multiple Choice Questions

1. What is the melting point of aluminium (in °C)?

(a) 250 (b) 478
(c) 660 (d) 728

2. Which of the following alloy can be used for explosion proof tools?

(a) Cu–Ni (b) Al–Mg
(c) Cu–Be (d) Fe–Cr

3. Inconel is an alloy that contains __________

(a) Ni and Cr (b) Ni, Cr and Fe
(c) Cr and Fe (d) Ni and Fe

4. Which of the following is the lightest structural metal?

(a) Magnesium (b) Aluminum
(c) Titanium (d) Copper

5. Which of the following is the costliest among the nonferrous materials?

(a) Magnesium (b) Aluminium
(c) Titanium (d) Copper

6. Major constituent of Duralumin alloy is ___________

(a) Copper (b) Nickel
(c) Iron (d) Aluminum

7. Alloy of Ni and Fe is termed as ___________

(a) Brass (b) Bronze
(c) Duralumin (d) Invar

8. Which of the following is the lightest metal among the following?

(a) Magnesium (b) Copper
(c) Titanium (d) Lead

9. Which of the following is the lightest and lowest melting point among the following?

(a) Tin (b) Aluminium
(c) Titanium (d) Copper

10. Bronze is an alloy of copper and ______________.

(a) Zinc (b) Tin
(c) Titanium (d) Nickel

11. Monel is an alloy that consists of nickel and ___________.
(a) Copper (b) Bronze
(c) Chromium (d) Iron

12. Nichrome is an alloy of
(a) Ni and Fe (b) Ni and Cr
(c) Ni and Cu (d) All

13. Which of the following are used as antifriction bearing alloys?
(a) Pb-based alloys (b) Sn-based alloys
(c) Ti-Ni alloy (d) All

14. Which of the following has the lowest corrosion resistance?
(a) Al alloys (b) Ti alloys
(c) Mg alloys (d) Mild steels

15. Which of the following is not a magnesium alloy?
(a) AZ91D (b) 6061
(c) ZK60 (d) Ti-6Al-4V

16. AZ31B alloy mainly consists of
(a) Mg, Zn, Sn (b) Mg, Al, Zn
(c) Mg, Th (d) Mg, Zn, Zr

17. Which of the following titanium alloys consists of both bcc and hcp phases?
(a) Ti-6Al-4Zr-1V (b) Ti-6Mn
(c) Ti-6Al-4V (d) Ti-3Al-13V-11Cr

18. The primary advantage of using bronze bearings is:
(a) Excellent heat dissipation (b) High strength
(c) High melting point (d) Self-lubricating properties

19. IS:6754-1972 and IS:6751-1972 deal with
(a) Lead babbits (b) Aluminium bearings
(c) Tin babbits (d) Porous metal powder oil-impregnated bearings

20. IS:3980-1982 deals with
(a) Lead babbits (b) Aluminium bearings
(c) Tin babbits (d) Porous metal powder oil-impregnated bearings

21. Which of the following is not Al based alloy?
(a) Duralumin (b) Hindalium
(c) Magnalium (d) Invar

22. Which of the combination of phase(s) has better ductility in case of Ti alloys
(a) α-Ti (b) β-Ti
(c) (α+β) Ti (d) None of these

23. Titanium shows ___________ crystal structure at room temperature.
(a) FCC (b) BCC
(c) HCP (d) Simple cubic

24. Which of the following nonferrous element does not show allotropic transformation?
    (a) Ni (b) Ti
    (c) Al (d) Mg

ANSWERS															
1.	(c)	2.	(c)	3.	(b)	4.	(a)	5.	(c)	6.	(d)	7.	(d)	8.	(a)
9.	(b)	10.	(b)	11.	(a)	12.	(b)	13.	(a)	14.	(c)	15.	(b)	16.	(b)
17.	(b)	18.	(d)	19.	(b)	20.	(d)	21.	(d)	22.	(b)	23.	(c)	24.	(a)

Short and Long Answer Type Questions

1. What are the outstanding properties of aluminium? Of copper? Of magnesium? Of titanium?
2. Draw cooling curves for the following alloys: Al–2% Cu, Al–20% Cu, Al–33% Cu, Al–0.5% Si, Al–5% Si, and Al–15% Si.
3. Explain the meaning of the digits in the following aluminium alloys specifications: 2104-O, 5056-T4, 5056-H18, 7075-T6, 6061- T4, 6063-H14.
4. Which Aluminium alloy would be the best for utensils? Why?
5. Give the composition, specific properties, and one application of the following alloys: Duralumin, Hindalium, Magnalium, Invar, *Chromel A, Inconel,* Grade 1–Copper base (Class A) bearing alloy, Grade 1–Copper base (Class B) bearing alloy, Muntz metal, Brass, Bronze, AZ31B, AZ91D, ZK60, Ti6Al4, and Babbits.
6. Differentiate between the terms bronze and brass.
7. Draw cooling curves for the following alloys: Cu–1% Zn,, Cu–10% Zn, Cu–10% Sn, and Cu–10% Ni.
8. What do you understand by season cracking and dezincification? How do we minimize both?
9. What are the advantages and disadvantages of magnesium?
10. Differentiate between the alpha and beta titanium alloys.
11. Draw the cooling curve of pure titanium.
12. Explain why the two-phase titanium alloys are stronger than single -phase alpha alloys.
13. State at least two applications for alpha alloys, alpha-beta alloys, and beta alloys.
14. Why is "Babbit" suitable for bearing application?
15. Explain the term seizure in case of Al-Si bearing alloys. How may it be minimized?

Numerical Problems

1. Write down the possible invariant reactions in case of Al-Si system. Determine the volume% of phases present just above and below the eutectic point in case of Al-5%Si.
2. Write down the possible invariant reactions in case of Al-Cu system. Determine the volume% of phases present just above and below the eutectic point in case of Al-20%Cu.
3. Estimate the volume% of phases present at room temperature in case of Al-20%Cu alloy and Al-4%Cu alloy.
4. Suggest the suitable alloying elements, along with their composition, to be used to get BCC and HCP crystal structures in case of titanium alloys.

PRACTICAL

Experiment title: Study the microstructure of brass.

Apparatus/ materials to be used: Brass sample, emery papers, optical microscope, polishing machine.

Objectives:

(a) Observe the microstructure of brass using optical micrograph and identify the phases present.

(b) determine the volume fraction of the phases present in the microstructure.

Theory:

Students can refer to ASM Handbook on copper and copper alloys for more detailed discussion on different microstructures for various compositions of copper alloys, and ASTM E562 to understand the test method for determination of volume fraction in the microstructure of a given metal using manual point count method.

Procedure:

Students can refer to ASTM E3-11 for standard guide for preparation of metallographic sample preparation. ASTM E407-07 can be also be referred to determine the etchant for brass and other copper-based alloys.

Result:

The student should show the microstructure of the brass obtained in the optical microscope along with the scale and magnification used.

Conclusion:

(a) The phases present in the microstructure of brass are ....

(b) The volume fraction of the different phases is ....

KNOW MORE

Students can refer to advanced concepts on calculation of phase diagrams (CALPHAD) for multicomponent systems using Thermocalc software.

REFERENCES AND SUGGESTED READINGS

1. V. Raghavan; Materials science and engineering- A first course, sixth edition 2015, Prentice Hall India Learning Private Limited.
2. William F. Smith and J. Hashemi; Foundations of Materials science and engineering, fourth international edition 2006, McGraw-Hill Education (Asia).
3. David A. Porter, Kenneth E. Easterling, and Mohamed Y. Sherif; Phase transformations in metals and alloys, third edition 2009, CRC Press.
4. F. C. Campbell; Elements of metallurgy and engineering alloys, 2011, ASM International.
5. R. Balasubramaniam; Callister's Materials Science and Engineering, 2nd edition 2014, Wiley India Pvt. Ltd.

Dynamic QR code for further reading

4 Failure Analysis and Testing of Materials

UNIT SPECIFICS

Through this unit the following aspects of failure analysis and testing of materials have been discussed:

- Significance of failure analysis and concept of fracture
- Ductile and brittle fracture modes and Griffith's theory of brittle fracture.
- Fatigue and creep and associated mechanisms.
- Destructive testing.
- Non-destructive testing.

The fundamental concepts of failure analysis and fracture have been incorporated in this text to impart the significance of the topics to the students. The different fracture modes due to fatigue and creep have been explained along with the associated mechanisms, factors affecting these failures and preventive measures. Various destructive and non-destructive tests to analyse the mechanical properties of the material that support the determination of the causes of failure along with their working principle which imparts practical knowledge to the students regarding the corresponding mechanical tests have also been incorporated.

Besides giving many multiple-choice questions as well as questions of short and long answer types marked in two categories by following lower and higher order of Bloom's taxonomy, a list of references and suggested readings are given in the unit so that one can go through them for further advanced knowledge.

This section also highlights the examples of some interesting facts, applications of the subject matter for industrial applications on a variety of aspects and/or our day-to-day real life, case study related to environmental, sustainability, social and ethical issues (whichever applicable), and finally inquisitiveness associated with the topics of the unit.

RATIONALE

This unit helps the students to have fundamental understanding about the different mechanical properties of a material and methods to perform these mechanical tests to understand the mechanical behavior of the material. The students have been exposed to the approach to undertake to perform failure analysis of any material and not only determine the cause of the failure but also propose preventive solutions so that such failures do not appear on similar components again in near future. The unit also explains the effects of fluctuating stresses and time dependent plastic deformation on a material and subsequently their failure due to fatigue and creep, respectively. Finally, the different destructive and non-destructive tests, corresponding methods to perform the tests and interpretation s of the results have been explained so that the students can have both theoretical and practical understanding of the topic.

PRE-REQUISITES

Basic idea about engineering materials.

UNIT OUTCOMES

List of outcomes of this unit is as follows:

U4-O1: To understand brittle and ductile fracture.

U4-O2: To have a basic understanding of fatigue and creep failure.

U4-O3: To have knowledge about different destructive tests.

U4-O4: To know about the various non-destructive test methods.

Unit-4 Outcomes	EXPECTED MAPPING WITH COURSE OUTCOMES (1- Weak Correlation; 2- Medium correlation; 3- Strong Correlation)				
	CO-1	CO-2	CO-3	CO-4	CO-5
U1-O1	3	2	2	3	3
U1-O2	3	2	2	3	1
U1-O3	3	1	1	3	2
U1-O4	3	1	1	3	3

4.1 INTRODUCTION TO FAILURE ANALYSIS

The failure of engineering components is always undesirable due to several reasons such as; threat to human lives, economic losses and breakdown in the operation. Thus, it is important to not only understand the reason for the failure but also provide preventive measures so that such failures do not occur in near future. Failure analysis is considered to be the evaluation of the root cause for the failure of the equipment and to identify the measures to be taken to avoid repeat of similar failures. The most common causes are (a) improper materials selection and processing, (b) inadequate design of the component or (c) its misuse. Also, damage can occur to structural parts during service, and regular inspection and repair are important for their safe design. Figure 4.1 illustrates a standard scientific model to define the problem-solving process during the root cause failure analysis.

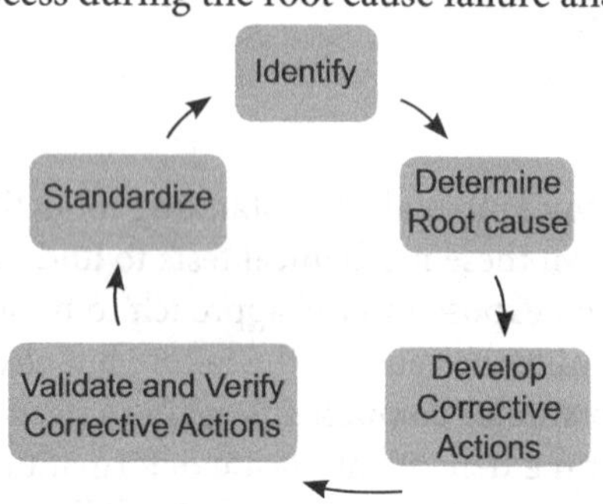

Fig. 4.1: Problem-solving model for root cause failure analysis

The major steps involved in the problem-solving model can be explained as;

Identify: Describe the current situation, define the deficiency of the failed component and its impact on the product line and collect data to provide a measurement of the deficiency.

Determine root cause: Analyze the problem and perform relevant tests to identify the causes.

Develop corrective actions: List possible solutions to mitigate and prevent recurrence of the problem. Generate alternatives and provide implementation plan.

Validate and verify corrective actions: Test corrective actions in pilot study. Measure effectiveness of change. Validate improvements.

Standardize: Incorporate the corrective action into the standards documentation system of the company to prevent recurrence in similar products or systems.

4.2 FRACTURE

Fracture refers to the failure of a material under load by breaking into two or more pieces. The applied stress may be tensile, compressive, shear or torsional. Fracture can occur under all service conditions. Materials subjected to alternating or cyclic loading fail due to fatigue. The fracture under such circumstances is called fatigue fracture. Materials used at elevated temperatures undergoes time dependent deformation and fail due to creep fracture. The characteristics of various fracture has been discussed extensively in this chapter.

Fracture can be classified into two major categories, namely, ductile fracture and brittle fracture. This classification is based on the ability of a material to experience plastic deformation (considerable distortion or degradation of the material before ultimate fracture). Ductile metals typically exhibit substantial plastic deformation with high energy absorption before fracture. However, there is little or no plastic deformation with low energy absorption accompanying a brittle fracture. Another important difference between the two fractures is related to the rate of propagation of crack. In brittle fracture, the rate of propagation of crack is very high. In fact, once the crack has nucleated, its propagation continues and ultimately material fractures without any increase in the magnitude of the applied stress. In ductile fracture, crack propagates slowly and only on increasing the magnitude of the applied stress. Fractures are also classified on the basis of crystallographic mode of fracture and the appearance of the fracture. Based on crystallographic mode of fracture, fractures are known as shear fracture or cleavage fracture, whereas on the basis of appearance, fractures are referred to as fibrous or granular fracture. A fibrous fracture is usually dull in appearance while a granular fracture is usually bright in appearance.

Any fracture process involves two steps namely; (a) crack formation and (b) propagation – in response to an imposed stress. The mode of fracture is highly dependent on the mechanism of crack propagation. Ductile fracture is characterized by extensive plastic deformation in the vicinity of an advancing crack. Furthermore, the process proceeds relatively slowly as the crack length is extended. Such a crack is often said to be stable as it resists any further extension unless there is an increase in the applied stress. However, for brittle fracture, cracks may spread extremely rapidly, with very little accompanying plastic deformation. Such cracks may be said to be unstable, and crack propagation, once started, continues spontaneously without an increase in magnitude of the applied stress.

Ductile fracture is almost always preferred to brittle fracture for two reasons: First, brittle fracture occurs suddenly and catastrophically without any warning as a result of spontaneous and rapid crack propagation. However, for ductile fracture, the presence of plastic deformation gives warning before failure, allowing preventive measures to be taken. Second, more strain energy is required to induce ductile fracture due to which these materials are generally tougher.

4.2.1 Ductile Fracture

Fracture with large amounts of plastic deformation is generally referred to as ductile fracture. Generally, all face centred cubic metals and their alloys exhibit ductile fracture under most of the service conditions. Failure of a material by ductile fracture can be attributed to the application of too high stresses. Many types of ductile fractures can occur when subjected to external stresses or put to use in service. A simple ductile fracture occurs when the material is subjected to uniaxial tensile stresses. Such a fracture proceeds by localized reduction in cross section in a part of the specimen under consideration. A ductile material, when subjected to tensile stresses, undergoes plastic deformation, which results in necking. Necking refers to the local deformation of a ductile material subjected to tensile loading (Figure 4.2(a)). Plastic deformation and necking proceed simultaneously.

During necking the cross-sectional area at the neck will reduce to such an extent that the necked portion will appear as a very thin sheet or as two small cones joined at the vertices. With such a reduced area, material under consideration will not be able to withstand stresses (load) and hence will fail by ductile fracture.

This common type of ductile fracture is the "cup and cone" type fracture (Figure 4.2(b)). It also occurs due to necking. On stressing a material, necking starts at a point of plastic instability. At this stage, the increase in the strength of the material due to plastic deformation (strain hardening) is unable to compensate for the stress requirement due to decrease in the cross-sectional area around the necked region. A triaxial stress concentration state gets induced here and many fine cavities originate within the central part of the necked region. These fine cavities coalesce to form voids. These voids continue to grow and finally result in the formation of a crack. Crack, so formed, has its long axis perpendicular to the direction of stress.

Once the crack has formed, it extends (grows) in a direction perpendicular to the axis of loading and tends to reach near to the external surface of the necked region. Propagation of such a crack to the outer surface of the necked region by shear deformation at an angle of about 45° with the tensile axis finally results in the fracture. Ductile fracture is characterized by appreciable amount of plastic deformation prior to and during the propagation of crack and by slow crack propagation. Ductile fracture normally occurs in a transgranular manner in metals that have good ductility and toughness. Figure 4.2(c) shows the presence of dimples present in the fractured surface of a low carbon steel.

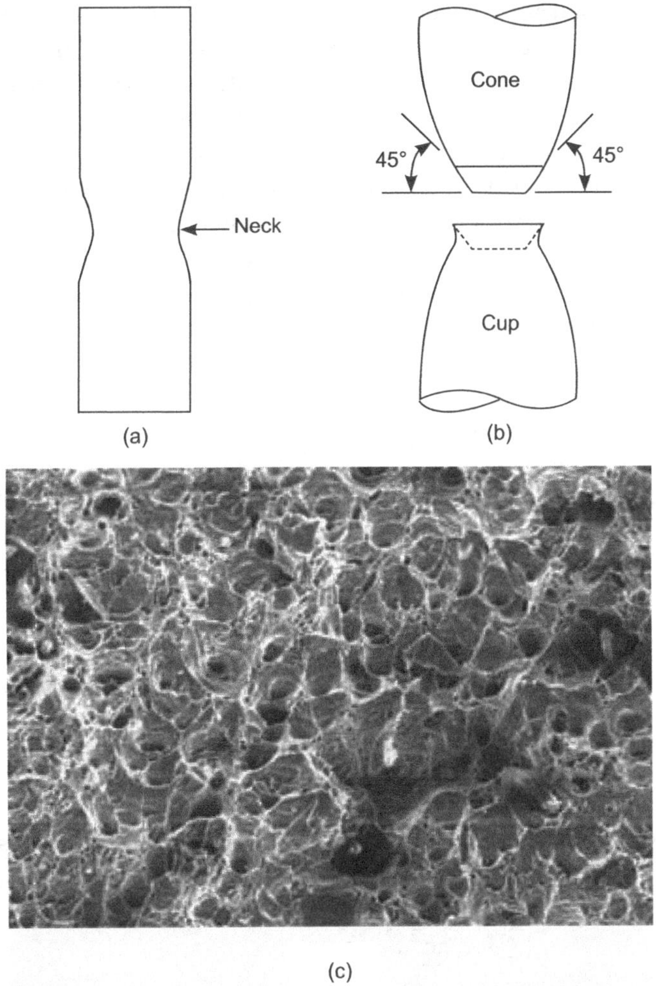

Fig. 4.2: (a) Neck formation at maximum tensile load in a ductile material, (b) Schematic representation of a ductile tensile failure characterising cup-cone like fracture, (c) Scanning electron micrograph of a low carbon steel exhibiting ductile fracture.

4.2.2 Brittle Fracture

Fracture that occurs after little or no plastic deformation is called brittle fracture. High strength metals and alloys having low ductility and toughness values are prone to failure by brittle fracture. White and grey cast irons fail by brittle fracture under all service conditions. Brittle fractures are usually observed under conditions of sudden loading rather than overloading. For most brittle crystalline materials, crack propagation corresponds to the successive and repeated breaking of atomic bonds along specific crystallographic planes and such a process is termed cleavage.

Griffith studied brittle fracture extensively and was the first to propose a mechanism for brittle fracture in highly brittle materials like glass. Hence, the mechanism of brittle fracture is most popularly referred to as Griffith theory of brittle fracture. According to Griffith, microcracks exist in all brittle materials which propagate and cause fracture. These pre-existing microcracks may be attributed to many flaws which usually occur during fabrication processes such as solidification, machining, welding, heat treating, forming or due to the interaction of dislocations with various imperfections/defects or particles of phases/impurities present in the material. Griffith assumed that the cracks are elliptical in shape. A very high concentration of stress gets developed at the tip of such cracks when the material is subjected to external stress. When the intensity of stress concentration exceeds the fracture strength at and around the tip of the crack, the crack propagates and ultimately causes fracture. Under tensile loading, the elastic energy stored per unit volume will increase with increase in load (stress). If a crack grows, i.e. increases in length, the growth is associated with an increase in surface energy as a result of increase in surface area. For an elliptical crack (Figure 4.3) of length 2a, the maximum stress (σmax) at the tip of the crack is given by Griffith equation as:

$$\sigma_{max} = 2\sigma\sqrt{\frac{a}{r}} \tag{4.1}$$

where σ is the applied tensile stress normal to crack length, and r is the radius of curvature at the crack tip.

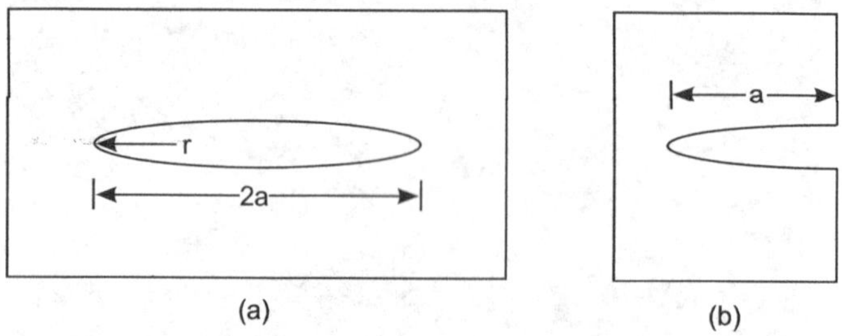

Fig. 4.3: Illustration of (a) An interior elliptical or lens shape crack, (b) A surface crack.

The elastic strain energy (UE) for unit volume is given by the relation:

$$U_E = \frac{\pi\sigma^2 a^2}{E} \tag{4.2}$$

where, E is Young's modulus

The surface energy (US) associated with such a crack of unit width can be expressed in terms of crack length as

$$U_S = 4\gamma_S a \tag{4.3}$$

where, γ_s is the surface energy per unit area.

As the length of the crack increases, surface energy associated with the crack also increases. However, elastic strain energy reduces with increase in the length of the crack. This reduction in elastic strain energy is due to the stress relieving of material close to the surfaces of the crack. Griffith postulated that loss of elastic strain energy should be equal to the increase in surface energy of the crack for the propagation of the crack.

Hence, according to Griffith criterion,

$$\frac{dU_E}{da} = \frac{dU_S}{da} \tag{4.4}$$

$$\sigma = \left[\frac{2\gamma_S E}{\pi a}\right]^{\frac{1}{2}} \quad (4.5)$$

The above equation shows that the tensile strength of a brittle material is largely governed by the length of the longest pre-existing crack in the material.

As stated earlier, Griffith suggested the mechanism of fracture for brittle materials. It is not useful for metallic materials. Cracks in such materials do not exhibit perfect elastic behaviour. In metallic materials, some plastic deformation, howsoever small, always precedes fracture. Therefore, even in brittle metallic materials there will be plastic deformation not only at the tip of the crack but also in the region adjacent to the crack tip. Orowan and Irwin modified the Eq. (4.5) by taking into account the energy required for plastic deformation as follows:

$$\sigma = \left[\frac{(2\gamma_S + \gamma_P)E}{\pi a}\right]^{\frac{1}{2}} \quad (4.6)$$

Where γ_P is the energy required for plastic deformation per unit area of the crack.

4.2.3 Fatigue

Fatigue corresponds to the behaviour of materials under the fluctuating or repeating cycles of stress or strain which cause deterioration (or reduction in strength and ductility) of the material that eventually results in a progressive fracture. The final fracture occurs in a brittle manner. Fatigue fracture occurs at a stress much lower than the tensile strength or even lower than the yield strength of the material. However, the local concentration of stresses, which do not reduce the static strength, may often lead to premature failure by fatigue. In fact, the number of failures in service by fatigue is much greater than static failures. The most unusual characteristic of fatigue failure is the lack of plastic deformation in the region of fractures, even in a metal like mild steel which behaves as ductile metal under static load. In this respect fatigue failure is similar to impact failure that occurs under dynamic loading conditions. Fatigue cracks are usually fine and difficult to detect. Once these cracks have grown to macroscopic size they may spread and cause complete failure in a short time without giving any warning. Thus, fatigue failure is catastrophic in nature and is one of the most common causes of fatigue in elements of many structures and machines.

Fatigue fractures are caused by the simultaneous action of cyclic stress, tensile stress and plastic strain. If any one of these three is absent, fatigue crack will not initiate and propagate. The cyclic stress initiates the crack, the tensile stress causes crack to grow through localized plastic strain. In general, the factors that favour fatigue failure to occur include a tensile stress of sufficiently high value, a large enough variation in applied stress, a sufficiently large number of cycles of stress, stress concentration, corrosive environment, overload and temperature. Fatigue property is an important consideration in the design of many components subjected to repeated cycles of stress. These components include motor shafts, bolts, springs (leaf as well as coil ones), turbine blades, gear teeth, connecting rod, pin, steam and gas turbine parts, rails, railway axles, car axles, wire ropes, etc. The number of fluctuating stress cycles may vary from millions to billions over the service life of these components.

4.2.3.1 Cyclic stresses

The applied stress may be axial (tension–compression), flexural (bending), or torsional (twisting) in nature. In general, three different fluctuating stress–time modes are possible. One is represented schematically by a regular and sinusoidal time dependence in Figure 4.4(a), where the amplitude is symmetrical about a mean zero stress level, for example, alternating from a maximum tensile stress (σmax) to a minimum compressive stress (σmin) of equal magnitude; this is referred to as a reversed stress cycle. Another type, termed a repeated stress cycle, is illustrated in Figure 4.4(b); the maxima and minima are asymmetrical relative to the zero-stress level. Finally, the stress level may vary randomly in amplitude and frequency, as exemplified in Figure 4.4(c).

Several parameters indicated in Figure 4.4(b) are used to characterize the fluctuating stress cycle. The stress amplitude alternates about a mean stress σ_m, defined as the average of the maximum and minimum stresses in the cycle, or

$$\sigma_m = \frac{\sigma_{max} + \sigma_{min}}{2} \tag{4.7}$$

The range of stress σ_r is the difference between σ_{max} and σ_{min}, namely,

$$\sigma_r = \sigma_{max} - \sigma_{min} \tag{4.8}$$

Stress amplitude σa is one-half of this range of stress, or

$$\sigma_a = \frac{\sigma_r}{2} = \frac{\sigma_{max} + \sigma_{min}}{2} \tag{4.9}$$

Finally, the *stress ratio R* is the ratio of minimum and maximum stress amplitudes:

$$R = \frac{\sigma_{min}}{\sigma_{max}} \tag{4.10}$$

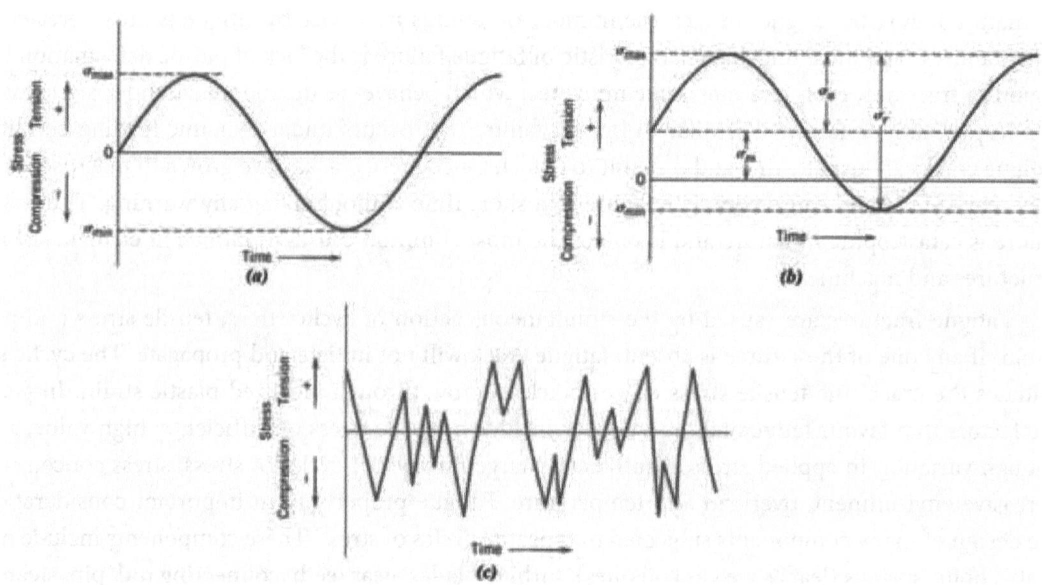

Fig. 4.4: Variation of stress with time that accounts for fatigue failures. (*a*) Reversed stress cycle, (*b*) Repeated stress cycle, and (*c*) Random stress cycle.

4.2.3.2 The S-N curve

Fatigue life is measured in terms of number of cycles of failure rather than in terms of time. Therefore, the number of cycles causing fracture for each specimen at a fixed stress is noted. The stress values (S) and the corresponding number of cycles of failure (N) are expressed in the form of a curve popularly known as S-N curve. As there may be many millions of stress cycles, a log scale is generally used for N. Figure 4.5 shows S-N curves for two different types of materials, one containing a notch and the other without a notch. It is apparent that the presence of a notch on the fatigue specimen greatly affects the S-N curve. A notch provides a stress concentration where a fatigue crack can initiate easily. The maximum stress at which a material fails due to repeated application of load for a specified number of cycles is termed as the fatigue strength. In other words, fatigue strength is the maximum alternating stress which a material can withstand without failure for a given number of cycles. Many of the materials exhibit a limiting stress value below which they can operate for an indefinitely large number of cycles without fracture. This limiting stress is commonly known as fatigue limit [Figure 4.5(a)]. It is often called as endurance limit which can be defined as the maximum alternating stress that a material can endure over a specified number of stress cycles without fracture [Figure 4.5(b)].

Thus, the condition of the surface is an important factor affecting the life of the material. Highly polished engineering surfaces show much higher fatigue life than the surfaces with scratches and roughness. Surface residual stresses greatly affect the fatigue life of the material. If the residual stresses are tensile in nature, the fatigue life is considerably diminished. On the other hand, if these stresses are compressive in nature, the fatigue life is much improved.

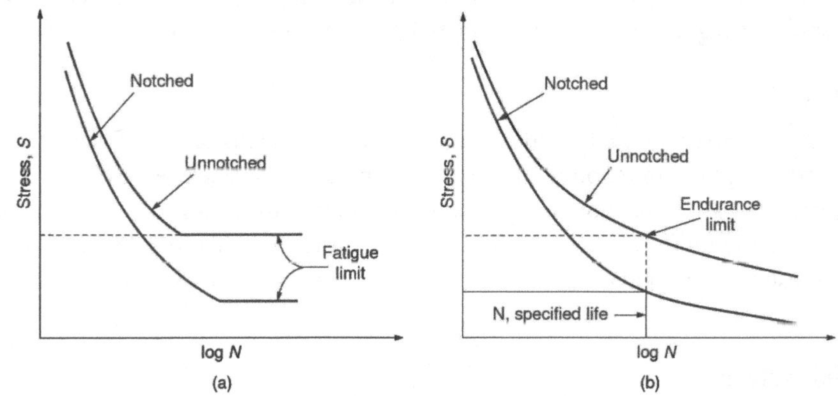

Fig. 4.5: Nature of the fatigue curves for notched and unnotched specimens. Number of stress cycles is plotted as log N. (a) Fatigue limit, (b) Endurance limit for a given number of stress cycles.

There are essentially two forms of fatigue failure, namely, low-cycle fatigue and high cycle fatigue. In low cycle fatigue, the maximum stress in any cycle is greater than the yield strength but less than the tensile strength of the material and the number of cycles of failure is low, generally less than 1000. In high cycle fatigue, as the name implies, the number of cycles of failure required is large (say in the range 105-106 cycles) and therefore the maximum stress in a cycle generally does not exceed the static yield strength of the material.

4.2.3.3 Factors affecting fatigue

The following are the factors that affect the fatigue property of metallic materials:

(a) Alloy composition
(b) Stress concentration
(c) Size of the specimen
(d) Surface condition
(e) Metallurgical structure

Alloy Composition

Those alloying elements which increase the tensile strength of a metal and hence the alloy are found to increase the fatigue strengths as well. Steels are the most widely used materials for applications involving cyclic stresses. Carbon in steels has been found to have the maximum effect on fatigue strength. Carbon increases the fatigue strength of steels not only through solid solution strengthening but also by strain ageing effect.

Stress Concentration

The points of stress raiser in the fatigue specimen greatly affect the fatigue strength. The points of stress raiser such as keyways, holes or notches on a component are the preferential sites for the initiation of a fatigue crack. In addition, the changes in the section size and surface irregularities such as machine markings, porosity, inclusions, decarburised regions, etc., also influence the fatigue life of a material to a considerable extent. A surface irregularity acts as a notch. The stress concentration at the root of the notch causes the gradient of stresses from the notched region towards the axis or centre of the specimen. Many of the surface irregularities can be eliminated by polishing the surface.

Size of the Specimen

If the stress cycle is simple reversal type, the stress changes from a tension peak to compression peak. The effect of such stress cycle is more on a thin section than in case of thick section. The distance between surface and centre (axis) will be more in thick section than in thin section. Thus, in a thicker section the gradient of stress or strain from surface to centre will be more than that in a thin specimen. In addition, increased size results in an increased surface area. Thus, there is more likelihood that a thicker section will fail.

Surface Characteristics

A fatigue crack is generally initiated at the surface, therefore, the nature of surface has great influence on the fatigue life of a component. One of the characteristics of the surface is its finishing. For the same material, variation in surface finishes will result in variation in fatigue strength. Smoother the surface, longer will be the fatigue life of the component. It has been observed that fatigue life increases with decrease in extent of surface roughness. In general, a buffed surface is regarded as a smooth surface, whereas an ordinary machined surface as rough. Another important surface characteristic of the material experiencing cyclic stress is the surface hardness. It has been observed that the fatigue strength of a material is greatly increased if its surface is made harder. A hard surface tends to resist initiation of a crack. Steels are case hardened to not only improve the surface hardness, and hence, the wear resistance, but also to improve fatigue life of the component. In contrast, electroplating, in general, results in lowering the fatigue strength of the metal.

Residual Stresses

The stresses that are left behind in an object after the process of manufacturing is finished are called residual stresses. Though these stresses are of very low magnitude, but may be dangerous when they are tensile in nature. Residual compressive stresses have been observed to be beneficial from the point of view of fatigue strength of the metal or alloy. It is therefore desirable to introduce compressive stresses in a component experiencing cyclic loading so that fatigue life of the component can be increased. Shot peening appears to be the best and most effective method of introducing compressive stresses in the surface. The other method is heat treatment at the appropriate stage in manufacture. For instance, heat treatment of gear after machining results in lowering or relieving the residual stresses. There are some heat treatments, such as carburizing which introduces compressive stresses at the surface of the component.

Metallurgical Structure

Among the metallurgical factors that affect the fatigue properties include microstructure, stacking fault energy, grain size, heat treatment, presence of soft surface spots caused by decarburization, formation of retained austenite or the formation of non-martensitic phase. The type of non-metallic inclusions and their orientation with respect to maximum tensile stress also affect the fatigue limit of the material.

Materials with high Stacking Fault Energy (SFE) allow dislocations to cross-glide easily past obstacles and lead to the formation of slip bands. These slip bands under alternating stress conditions turn into intrusions or extrusions which are the centres for nucleation of a fatigue crack. Materials with lower SFE, on the other hand, make cross slip difficult and dislocations are restricted to move in a planar fashion to a limited extend. A limited local concentration of plastic deformation delays the formation of intrusions or extrusions, and hence, the nucleation of fatigue crack will be delayed. A low SFE coupled with a fine grain structure also results in improvement in fatigue properties. Here, grain boundaries control the rate of cracking.

In steels, the type of heat treatment affects the fatigue properties. In quenched and tempered steels, the fatigue limit is found to improve as the tempering temperature is lowered to obtain hardness in the range Rc 45 to Rc 55, depending on the kind of steel. When the quenched and tempered microstructure of steel is compared with austempered structure for the same hardness it is found that the latter imparts much superior fatigue properties than the former.

4.2.4 Creep

Creep is the permanent deformation of a material under load as a function of time. It is appreciable only at temperatures above 0.4Tm (Tm = melting temperature of respective metals/alloys). Room temperature for iron is 0.16Tm and for copper it is 0.22Tm so that creep at room temperature is negligible in these materials. On the other hand, room temperature is about half of the melting point (in kelvin) for lead and so it undergoes creep at room temperature.

The most important properties used in the design of materials for elevated temperature applications are the creep strength and creep rupture strength. Creep strength (also commonly called creep limit) is defined as the stress that a material can withstand for a specified period of time without excessive deformation. The creep rupture strength or simply the rupture strength is defined as the limiting stress that a material can sustain for a specified period of time without rupture. In an alternative way, creep strength can also be defined as the constant stress that will produce a specified magnitude of strain (or creep) in a material over a given period of time at a constant temperature.

4.2.4.1 Creep curve

Creep curve is obtained by placing a load on a standard specimen at constant temperature and the deformation produced with time is measured. Figure 4.6 is a schematic representation of the typical constant-load creep behavior of metals. Upon application of the load, there is an instantaneous deformation, as indicated in the figure, that is totally elastic. The resulting creep curve consists of three regions, each of which has its own distinctive strain–time feature. Primary or transient creep occurs first, typified by a continuously decreasing creep rate—that is, the slope of the curve decreases with time. This suggests that the material is experiencing an increase in creep resistance or strain hardening. For secondary creep, sometimes termed steady-state creep, the rate is constant—that is, the plot becomes linear. This is often the stage of creep that is of the longest duration.

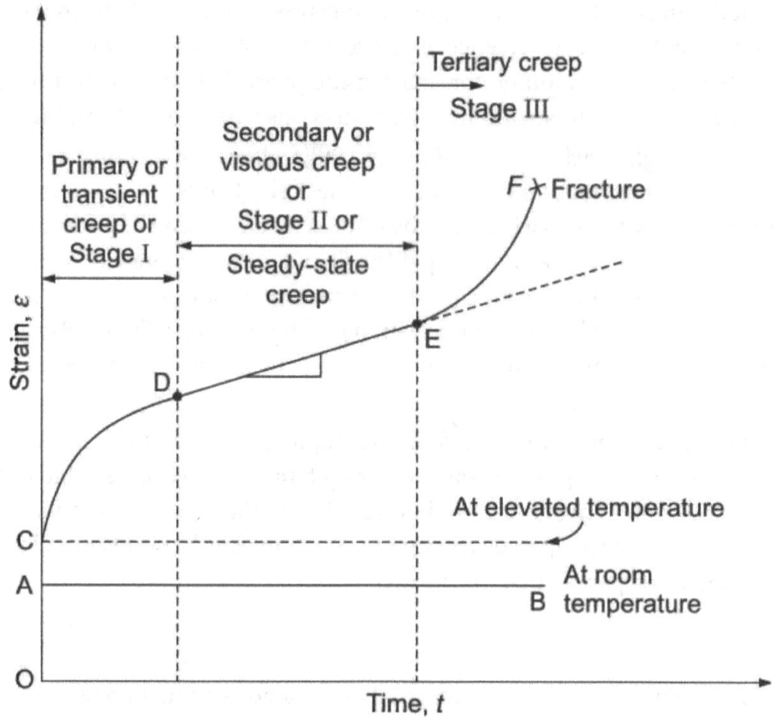

Fig. 4.6: An ideal creep curve showing three stages of creep.

The constancy of creep rate is explained on the basis of a balance between the competing processes of strain hardening and recovery, where recovery being the process by which a material becomes softer and retains its ability to experience deformation. Finally, for tertiary creep, there is an acceleration of the rate and ultimate failure. This failure is frequently termed rupture and results from microstructural and/ or metallurgical changes—for example, grain boundary separation, and the formation of internal cracks, cavities, and voids. Also, for tensile loads, a neck may form at some point within the deformation region. These all lead to a decrease in the effective cross-sectional area and an increase in strain rate.

4.3 DESTRUCTIVE TESTING

4.3.1 Tensile Testing

A stress-strain curve is obtained from a tensile test. An ordinary tensile test is conducted at or near room temperature under the uniaxial tensile load that is applied on the specimen at slow strain rate (~ 10-3 s-1) . A standard specimen of specified geometry is used for tensile testing as shown in Figure 4.7. The specimen is symmetric about its long axis so that the load is distributed uniformly over its cross-section. During the test, load is applied uniaxially along the specimen length and increases gradually until fracture occurs. The elongation produced is measured at frequent intervals. The load and elongation data are plotted as a curve called load-elongation curve. Since the loads will differ for different sections of the specimen and likewise the elongation will vary with gauge length, it is important to remove the size effects so that the produced data is characteristic of a given material rather than that of the particular specimen. This can be done by dividing the loads and elongations by constant original dimensions. The load (P) and elongation data are thus expressed as average stress and average strain by dividing them by original area of cross section (A0) and gauge length (L0) of the specimen, respectively. That is,

$$\text{Average stress } (\sigma) = \frac{\text{Load}}{\text{original area of crossection}} \tag{4.11}$$

$$\sigma = \frac{P}{A_0} Kgmm^{-2} \tag{4.12}$$

$$\text{Average strain } (\varepsilon) = \frac{\text{change in length}}{\text{original gauge length}} = \frac{\Delta L}{L_0} \tag{4.13}$$

$$\varepsilon = \frac{L - L_0}{L_0} \tag{4.14}$$

where, L is the instantaneous length. The average stress and average strain data so obtained are plotted in the form of stress-strain curve. Since both load and elongation are divided by constant values, namely, original cross-sectional area and gauge length, the stress-strain curve is commonly called Engineering, conventional or nominal stress strain curve.

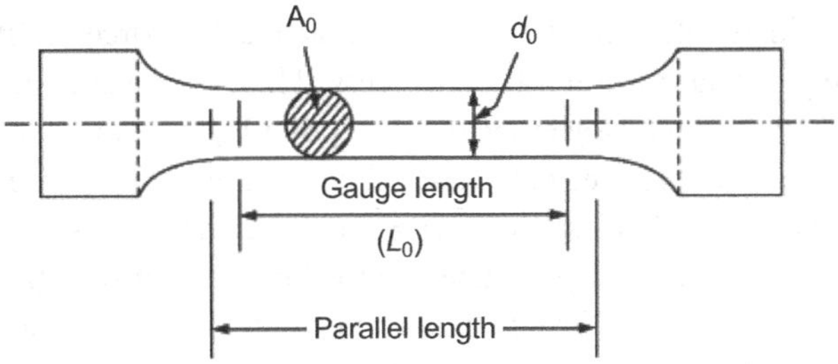

Fig. 4.7: Geometry of tensile specimen

A typical stress-strain curve of a mild steel is shown in Figure 4.8. In the range OA, the strain produced varies in proportion to the applied stress. The curve in this range follows the law of proportionality commonly called Hooke's law. The curve is linear in this range. The constant of proportionality is the ratio of stress to strain and is called Young's modulus or the modulus of elasticity. The point A represents the maximum stress up to which stress is proportional to strain and is referred to as Proportional limit. In order to determine proportional limit a very sensitive extensometer is to be used so that it can detect even a slightest deviation from straight line. All materials do not necessarily exhibit a linear relationship between stress and strain, though they display elastic behaviour. A typical example of one such material is 'rubber' which is an elastic material but does not exhibit a linear stress-strain relationship at all. Other materials which exhibit this type of behaviour include copper, cast iron and sometimes aluminium.

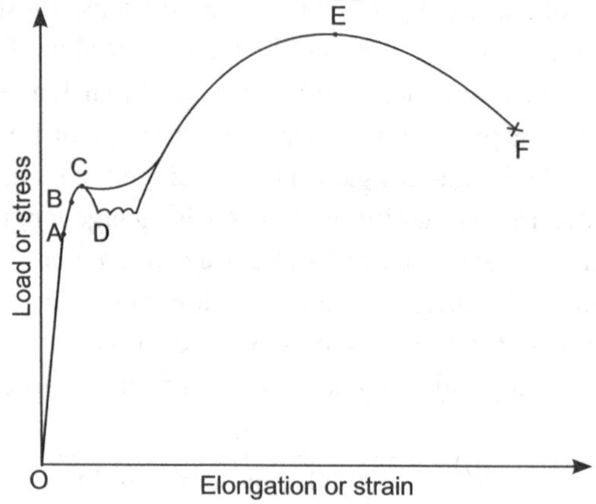

Fig. 4.8: Typical engineering stress-engineering strain curve for mild steel,

As the stress increases beyond the point A, the curve begins to deviate from linearity. However, the specimen still exhibits elastic behaviour up to a stress that corresponds to the point B, from where if the stress is removed, the specimen will regain the original dimensions. Many materials show this type of behaviour. The point B is called elastic limit and is defined as the maximum stress up to which the specimen behaves elastically. As the stress exceeds the elastic limit, the specimen begins to yield, i.e. deforms permanently. Certain metals and alloys, in particular those with BCC structure and which contain a small amount of interstitial solute element, are characterised by a sharp yield point. A typical example of such materials is mild steel. This point (point C in Figure 4.8) at which marked increase in strain occurs without a simultaneous increase in stress is called the yield point. The stress corresponding to this point is called yield strength. In other words, it represents the resistance of the material against plastic deformation. If yielding occurs with a drop in stress, as illustrated in Figure 4.8, the point C is called the upper yield point. Once this point is reached the force resisting deformation decreases and attains a relatively constant lower value of stress. This lower value of stress at which appreciable plastic deformation occurs is called lower yield point (point D). Many materials can tolerate small strains beyond the elastic limit without appreciable plastic deformation. These strains are typically 0.1 % or 0.2% of the original dimensions. Strains less than 0.1% are difficult to measure with accuracy. The stress at which a material exhibits a specified permanent set is known as proof stress. This stress is referred to as the yield strength of the material. Proof stress can be determined by drawing a line through 0.1 % or 0.2% of strain as offset along the abscissa and parallel to the straight line portion of the tensile curve. The intersection of this line with the tensile curve gives a stress called yield strength or proof stress.

Beyond the yield point, deformation occurs with continuous increase of stress but to a relatively small amount until the maximum point E is reached. The nature of the stress-strain curve is parabolic in this range which is associated with homogeneous plastic deformation processes such as irreversible movement of dislocations in metals. Thereafter, further strain occurs with a continuous decrease in stress until the specimen fractures (point F). The stress at E, is therefore, the maximum stress before the occurrence of fracture and is known as ultimate tensile strength (UTS) or simply tensile strength. Tensile strength of a material is therefore defined as its resistance against fracture.

Ductility is another important mechanical property which is a measure of the degree of plastic deformation that has been sustained at fracture. Ductility may be expressed either as precent elongation (%EL) or percent reduction in area (%RA). The percent elongation is the percentage of plastic strain at fracture and could be expressed as;

$$\%EL=\left(\frac{l_f-l_0}{l_0}\right)\times 100 \tag{4.15}$$

where l_f is the final length (It is the length between the gage marks after testing when the fractured surface is fitted together) and l_0 is the initial length (It is length between the gage marks on the specimen before tensile testing).

The percentage reduction in area could be expressed as;

$$\%RA=\left(\frac{A_0-A_f}{A_0}\right)\times 100 \tag{4.16}$$

Where A_0 is the initial cross-sectional area and A_f is the final cross-sectional area at the point of fracture.

Most of the ductile metals show a drop in load after ultimate tensile point. This is because of necking that occurs within the gauge length of the specimen at maximum stress. This nature of the curve is because of the two opposing effects. The first effect is due to the strain hardening that result due to multiplication of dislocations during plastic deformation and consequently more stress is required for further plastic deformation. The second effect is related to the elongation of the specimen which results in the decrease in its diameter. As the diameter decreases, the load carrying capacity of the specimen must decrease. Initially strain hardening predominates over the decrease in load carrying capability of the specimen and therefore overall stress required for plastic deformation increases until a maximum stress where onset of necking occurs. After this, decrease in diameter of the specimen is rapid; resulting in the second factor to predominate, i.e. the decrease in load carrying capability of the specimen is much faster than strain hardening. Decrease in localised diameter of the specimen needs lower value of load for further deformation. Hence there is decrease in stress after onset of necking until fracture occurs (point F).

Problem 4.1: The tensile stress-strain curve for a non-ferrous specimen is shown in Figure below. Determine the following.

(a) the modulus of elasticity; (b) the yield strength at a strain offset of 0.002; (c) the maximum load that can be sustained by a cylindrical specimen having an original diameter of 14.5 mm; (d) the change in length of a specimen originally 230 mm long that is subjected to a tensile stress of 320 MPa.

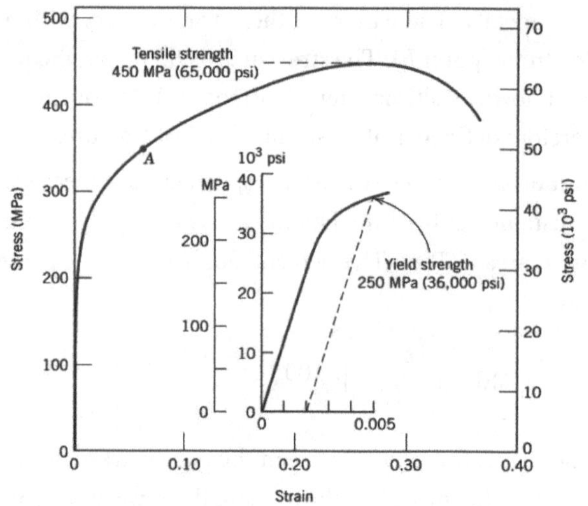

Fig. 4.9: The stress-strain curve for a non-ferrous specimen

Solution:

(a) The modulus of elasticity is the slope of the elastic or initial linear portion of the stress strain curve. In mathematical terms,

$$E = \text{slope} = \frac{\Delta\sigma}{\Delta\varepsilon} = \frac{\sigma_2 - \sigma_1}{\varepsilon_2 - \varepsilon_1}$$

$$= \frac{(150-0)}{(0.0016-0)} = 93.8\ GPa$$

(b) The 0.002 strain offset line is constructed as shown in the inset; its inter-section with the stress–strain curve is at approximately 250 MPa, which is the yield strength of the brass.

(c) The maximum load that can be sustained by the specimen is calculated as;

$$F = \sigma A_0 = \sigma\left(\frac{d_0}{2}\right)^2 \pi$$

$$= (450\times10^6)\left(\frac{12.8\times10^{-3}}{2}\right)^2 \pi = 57{,}900\,N$$

(d) To compute the change in length, Δl, it is first necessary to determine the strain that is produced by a stress of 345 MPa. This is accomplished by locating the stress point on the stress–strain curve, point A, and reading the corresponding strain from the strain axis, which is approximately 0.06. For $l_0 = 250$ mm, we have

$$\Delta l = \varepsilon l_0 = (0.06)(250) = 15\text{mm}$$

4.3.2 Hardness Testing

Hardness is defined as the property of the material to resist permanent or plastic deformation. There are three general types of hardness measurements depending on the manner in which the test is conducted.

These are (a) scratch hardness, (b) indentation hardness and (c) rebound, or dynamic hardness. The present chapter deals only with the indentation hardness as it is of major engineering interest for metals and alloys.

4.3.2.1 Brinell hardness test

The first widely accepted standardized hardness test was first proposed by J.A. Brinell in 1900. This method of measuring the hardness is based on the measurement of size of impression made by a hardened steel ball of 10 mm diameter under a known static load (usually 3000 kgf). The Brinell hardness (HB) is the ratio of load *P* to the surface area of indentation and is expressed quantitatively in number (without unit) known as Brinell hardness number:

$$\text{Brinell hardness number (BHN)}= \frac{\text{load}}{\text{area of indentation}} = \frac{P}{A} \tag{4.17}$$

Here *A*, is the area of contact between the ball and indentation. If deformation of the ball during indentation is neglected, the surface of contact is a spherical segment, whose area is given as:

$$\text{A} = \frac{t}{D} \neq D^2 \tag{4.18}$$

$$\text{A} = \pi \text{Dt} \tag{4.19}$$

where, *D* is the diameter of indenter and *t* is the depth of impression. Since the depth *t* can only be measured with great difficulty therefore diameter *d* of impression is measured instead of *t*. The diameter of impression should be the average of the two readings measured at right angles. In standard Brinell testing therefore diameter of indentation is measured. From the geometry shown in Figure 4.9, we get 't' as

$$t = \frac{D - \sqrt{D^2 - d^2}}{2} \tag{4.20}$$

Substituting the value of *t* in Eq. (4.19), Brinell hardness is given as:

$$\text{BHN} = \frac{2P}{\pi D(D - \sqrt{D^2 - d^2})} \text{ Kgf mm}^{-2} \tag{4.21}$$

In order to obtain the same (reliable) BHN values with a nonstandard load or ball diameter it is necessary to produce geometrically similar indentations. For indentations to be geometrically similar, the included angle subtended by the indentation edge at the centre of the ball (Figure 4.10) should remain constant. Thus, the basic principle involved is that when indenters of different diameter are used, the load to diameter must be varied in the ratio;

$$\frac{P_1}{D_1^2} = \frac{P_2}{D_2^2} = \frac{P_3}{D_3^2} \tag{4.22}$$

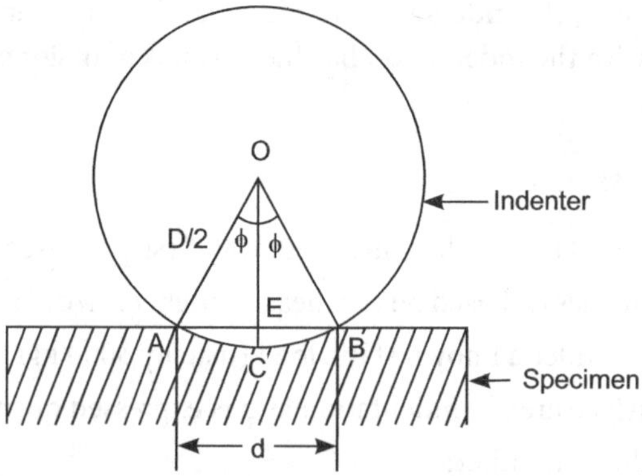

Fig. 4.10: Determination of area of spherical indentation ACB. The angle subtended by the edge of the indentation at the centre 'O' of the indenter is 2ϕ.

Unless precautions are taken to maintain the P/D2 constant, the BHN generally will vary with load.

The relatively large size of the Brinell impression could be an advantage in averaging out local heterogeneities. Moreover, the Brinell test is less influenced by surface scratches and roughness unlike other hardness tests. On the other hand, large size of impression is not desirable in certain components especially for the smaller and thin components. Moreover, deep and large indentations spoil the surface of the finished product. Hardened steels and surface hardened steel components cannot be tested by this method because of the limited range of hardness that can be measured by this method due to fear of flattening of indenter. The sample surface should be flat and there should not be any movement of sample in the horizontal direction during the hardness testing.

4.3.2.2 Rockwell hardness test

The Rockwell hardness test is the most widely used test in engineering industries, in particular, where mass production is carried out. This method directly gives the value of hardness indicated on the machine dial, and therefore, it is widely used for routine and quick checking of hardness. This test utilizes the depth of indentation, under constant load, as a measure of hardness. A minor load of 10 kgf is first applied to seat the specimen. This minimizes the amount of surface preparation needed and reduces the tendency for ridging or sinking in by the indenter. The major loads used are 60, 100 or 150 kgf depending on the geometry of the indenter and the hardness of the material under test. In Rockwell hardness test, two types of indenters are in use. The first being a hardened steel ball of 1/16 inch (1.58 mm) diameter and the other is a diamond cone indenter with its apex angle equals 120° having the rounded tip of radius 0.2 mm. It is also commercially known as Brale indenter.

Since the Rockwell hardness is dependent on the load and indenter, it is necessary to specify the combination which is used. This is done by prefixing the hardness number with a letter indicating the particular combination of load and indenter for the hardness scale employed. Hardened steel is tested on the C scale with the diamond indenter and a 150-kg major load. The useful range for this scale is from about Rc 20 to Rc 70. Softer materials are usually tested on the B scale with a 1.6 mm-diameter steel ball and a 100-kg major load. The range of this scale is from RB 0 to RB 100. The A scale (diamond penetrator, 60-kg major load) provides the most extended Rockwell hardness scale, which is usable for materials from annealed brass to cemented carbides.

The test is very rapid for measurement of hardness and is widely used for routine work and in engineering industries where the test is to be carried out on large number of pieces. Moreover, relatively lower load and pointed indenter causes no destruction to the surface of the component.

4.3.3 Bend Test

There are many structural applications in which the material is subjected to bending loads. In such cases, bending or flexural test rather than tensile test are more common to serve as a direct means of evaluating the behaviour of materials under bending loads. For brittle materials, tensile test is uncommon because of the difficulty in fabricating the sample of specified geometry and the difficulty encountered while the test is performed. Bend test is widely used for brittle materials such as glass, ceramics and concrete. In the test, the specimen is supported on two points. The distance between the supports must be large with respect to the other dimensions. If L is the length between the supports and t is the thickness, then L generally varies from 6t to 12t. The specimen geometry for rectangular section specimen is shown in Figure 4.10. The flexural strength (also sometimes called modulus of rupture) can be calculated by using the following expressions:

$$\text{Flexural strength} = \left(\frac{3}{2}\right)\left(\frac{PL}{bt^2}\right) \text{ for rectangular specimen of width b and thickness t.}$$

$$\text{Flexural strength} = 8\left(\frac{PL}{\pi D^2}\right) \text{ for cylindrical specimen of diameter D}$$

where P is the load at failure.

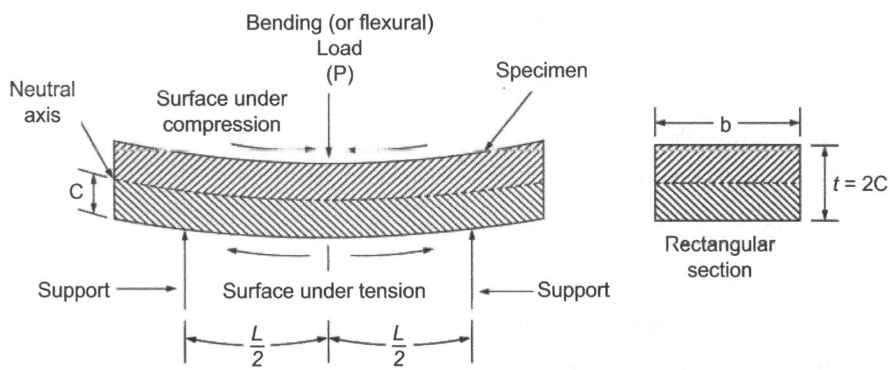

Fig. 4.11: Schematic representation for bend test. The specimen used is supported beam like.

4.3.4 Torsion Test

Torsion tests are made on materials to determine properties such as modulus of elasticity in shear, torsional yield strength, and modulus of rupture. Torsion tests could also be performed on full-sized parts, such as shafts, axles, and twist drills, which are subjected to torsional loading in service. It is frequently used for testing brittle materials, such as tool steels and is also used in the form of high temperature twist test to evaluate the forgeability of materials.

Torsion testing equipment consists of a twisting head, with a chuck for gripping the specimen and for applying the twisting moment to the specimen, and a weighing head which grips the other end of the specimen and measures the twisting moment or torque. The deformation of the specimen is measured by a twist-measuring device called a troptometer. The angular displacement of a point near one end of the test section of the specimen is determined with respect to a point on the same longitudinal element at the opposite end. A torsion specimen generally has a circular cross section as it represents the simplest geometry for the calculation of the stress. Since in the elastic range the shear stress varies linearly from a value of zero at the center of the bar to a maximum value at the surface, it is frequently desirable to test a thin-walled tubular specimen. This results in a nearly uniform shear stress over the cross section of the specimen.

Consider a cylindrical bar which is subjected to a torsional moment at one end (Figure 4.11). The twisting moment is resisted by shear stresses set up in the cross section of the bar. The shear stress is zero at the center of the bar and increases linearly with the radius. Equating the twisting moment to the internal resisting moment,

$$\mathrm{M_T} = \int_{r=0}^{r=a} rrdA = \int_0^a r^2 dA \tag{4.23}$$

But $\int_0^a r^2 dA$ is the polar moment of inertia of the area with respect to the axis of the bar. Thus,

$$\mathrm{M_T} = \frac{\tau J}{r} \tag{4.24}$$

$$\tau = \frac{M_T r}{J} \tag{4.25}$$

Where τ = shear stress, Pa

$\mathrm{M_T}$ = torsional moment, Nm

r = radial distance measured from the centre of bar, m

J = polar moment of inertia, $\mathrm{m^4}$

Since the shear stress is a maximum at the surface of the bar, for a solid cylindrical specimen where $\mathrm{J} = \pi \mathrm{D}^4/32$, the maximum shear stress is

$$\tau_{\max} = \frac{\frac{M_T D}{2}}{\frac{\pi D^4}{32}} = \frac{16 M_T}{\pi D^3} \tag{4.26}$$

For a tubular specimen, the shear stress on the outer surface is

$$\tau = \frac{16 M_T D_1}{\pi (D_1^4 - D_2^4)} \tag{4.27}$$

Where $\mathrm{D_1}$ = outside diameter of the tube and

$\mathrm{D_2}$ = inside diameter of the tube.

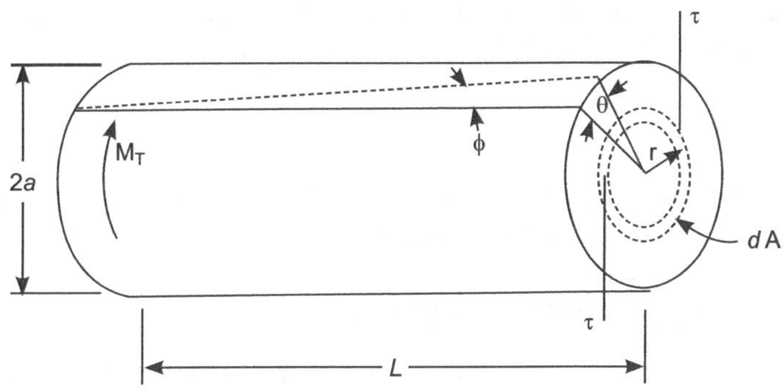

Fig. 4.12: Torsion of a solid bar

The troptometer is used to determine the angle of twist θ, expressed in radians. If L is the test length of the specimen, the shear strain is given by

$$\gamma = \tan\varphi = \frac{r\theta}{L} \tag{4.28}$$

4.3.5 Fatigue Test

When a component is subjected to many repetitive cycles of alternating stress, the breakage of the part is said to be a fatigue failure. A typical example of fatigue failure is the fracture of a wire when it is subjected to bending to and fro many times. In this case, a load is applied and removed repeatedly. Many service components such as gears, connecting rods, springs, steel paper clips, car axles, aircraft fuselage, etc. are subjected to fatigue stress cycles.

One of the methods of testing fatigue is to stress a rotating specimen in bending by stationary weights. This generates alternating tensile and compressive stresses in the specimen. For a given stress the number of cycles of failure is noted. The test is repeated for various stresses. From the results so obtained a graph is plotted between stress (S) and the number of cycles (N) of failure. This curve is commonly known as S-N curve. The type of machine used to generate alternating stresses on the specimen is called Rotating bending machine. In this machine, a standard fatigue specimen of circular cross-section (Figure 4.12) is loaded at a single point with transverse loads in a fixed plane while being rotated about its axis with respect to this plane. The other end of the specimen is fixed in the machine chuck. The load is applied through ball bearing. The transverse load sets up bending stresses in the specimen. When the load is applied at a single point [Figure 4.12(a)], the bending moment increases with increasing distance from the applied load point and precipitates at the base of the fillet at the end of the gauge section, i.e. near the fixed end. In effect, this represents a notched fatigue test, since the results will depend strongly on fillet geometry. As the specimen rotates, it passes through alternate compressive and tensile stresses as shown in Figure 4.12(b) and the number of the cycles of stress is equal to the number of revolutions of the machine. The stress σt at the tension side of the specimen is given as:

$$\sigma_t = \frac{32M}{\pi d^3} = \frac{32WL}{\pi d^3} \tag{4.29}$$

where M is the bending moment, W is the load on the specimen in kgf and L is the length of the specimen between the fillet and the point of loading and d is the minimum diameter of the specimen. As the specimen rotates and completes one half tum, the tensile side now comes under compression. The compressive stress σc is given as:

$$\sigma_c = -\frac{32WL}{\pi d^3} \tag{4.30}$$

The test is continued until specimen fractures. The number of stress cycles of failure is read out on the revolution counter attached with the machine.

4.4 NON-DESTRUCTIVE TESTING

The mechanical tests discussed in the previous section are destructive in nature and there is no assurance that the part used in the structure or machine will behave in the same way during service as it was tested. They may perform differently because other flaws might have been introduced during processing (such as heat treatment, machining or rolling) or fabrication (such as casting or welding). Some common flaws found in metallic products are seams, subsurface flaws and cracks resulting from quenching, embrittlement and fatigue. There is, therefore, a need to have methods by which flaws in the finished products can be detected or examined without affecting their performance. Non-Destructive Testing (NDT) is one such class of techniques widely used in industries. The basic objective of NDT of materials and structural components is the assessment of their probability of survival under known service conditions. The assessment is made through detection of defects or disorders that may lead to premature failure of the parts. Although NDT is basically understood to be an inspection procedure for detection of defects, yet the techniques of NDT can be applied for other kinds of qualitative analysis, like thickness measurement, assessment of surface finish, structural analysis and related physical properties like grain size, grain orientation, elastic constants, etc. The different categories of NDT are as follows:

(a) Visual inspection
(b) Magnetic particle inspection
(c) Liquid penetrant test
(d) Ultrasonic inspection
(e) Radiography

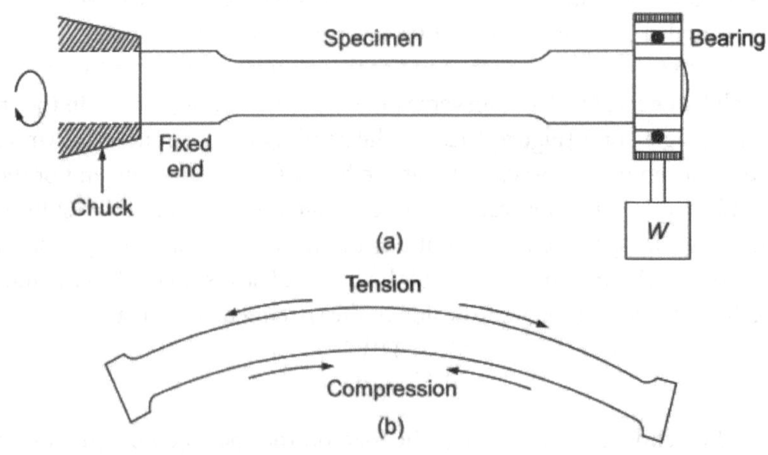

Fig. 4.13: Arrangement of fatigue specimen in the testing machine.

4.4.1 Visual Inspection

Visual inspection of a component is the cheapest and fastest method of NDT and is probably the most widely used among all NDT methods, in particular, when surface flaws are to be detected. Visual inspection can reveal surface defects or flaws such as weld cracks, weld seams, weld undercutting, surface porosity of welds or castings, etc. The basic principle involved in visual inspection consists of illuminating the object or the test specimen with light in the visible region. The specimen is then examined with the aid of eye with magnifying glass, and microscope or by light sensitive device such as a photocell. Unfortunately, the eye is not an accurate device and vision can vary from individual to individual. The eye is not reliable when differentiating between light intensities. Thus, for best results of a visual inspection, the contrast between the defect and the background must be good. The minimum size of defect that can be resolved with unaided eye also depends on this contrast. To achieve this contrast etching is done sometimes. Other physical aids used for visual inspection include magnifying glass, magnifying mirror, borescope (an instrument to inspect the inside of a narrow tube, bore or chamber) and endoscope (an instrument similar to borescope, but much easier to use). Applications of visual inspection can be summarised as follows:

(a) General inspection of the component for the presence or absence of corrosion or erosion product or cracking due to them
(b) Inspection of thin-walled tubular component for leakage
(c) Defects such as seams, tear cracks, porosity, mismatch, gross surface cracks, etc. in the weldments
(d) Minute discontinuities in pumps, compressors, turbo generator parts, instruments, etc.

Today, artificial intelligence (AI) is also used to facilitate automation in visual inspection process. It is also known as intelligent visual inspection. In this case, a computer is trained to recognise images and determine when they meet acceptable standards by adopting training, validation, and testing. The intelligent visual inspection process is faster, more accurate and cost-effective across a wide range of environments including hazardous areas. Today's car manufacturing companies use images and deep learning to identify defects quickly and consistently in the production process. Further, this method can be extended to identify corrosion occurring on the tops of wind turbines and to determine the faulty connectors within electronic components.

4.4.2 Magnetic Particle Inspection

The Magnetic particle inspection (MPI) technique works only on materials that can be easily magnetized (i.e. the ferromagnetic materials) and can detect discontinuities which are generally open to the surface. MPI is basically a combination of visual inspection and magnetic flux leakage testing. A surface crack of width as fine as 1 μm can be detected using best conditions of MPI. Subsurface flaws, that is, the flaws which lie just below the surface can also be revealed by this method. Ferromagnetic materials include iron, nickel, cobalt and their alloys and many precipitation hardening steels such as 17-4 PH, 17-7 PH and 15-4 PH stainless steels in the aged condition. These materials lose their ferromagnetism above the Curie point. Although this temperature is different for different materials, the Curie point for most ferro-magnetic materials is about 760°C. Non-ferromagnetic materials such as austenitic stainless steels, aluminium, copper, magnesium, titanium and their alloys cannot be inspected by this technique.

When a ferromagnetic component is magnetized, the magnetic lines of force (called magnetic flux) run predominantly inside the material. If the component contains a flaw such as a crack or an inclusion, it interferes with the magnetic lines of force. Some of these lines of force must exit at one edge and re-enter the other edge of the crack that exists in the component. That is, there is localised leakage of magnetic flux (Figure 4.13). Leakage is appreciable at location A (surface crack) while it is fuzzy at location B (sub-surface crack). Discontinuity at location C is undetectable as it is deeply seated within the workpiece. Even if the defect lies on the surface, it cannot be detected effectively, as it is oriented parallel to the field applied. The points of exit and re-entry at the edges of the discontinuity, form opposite magnetic poles. When minute magnetic particles are sprinkled over the surface of the specimen, part of these are attracted by these magnetic poles.

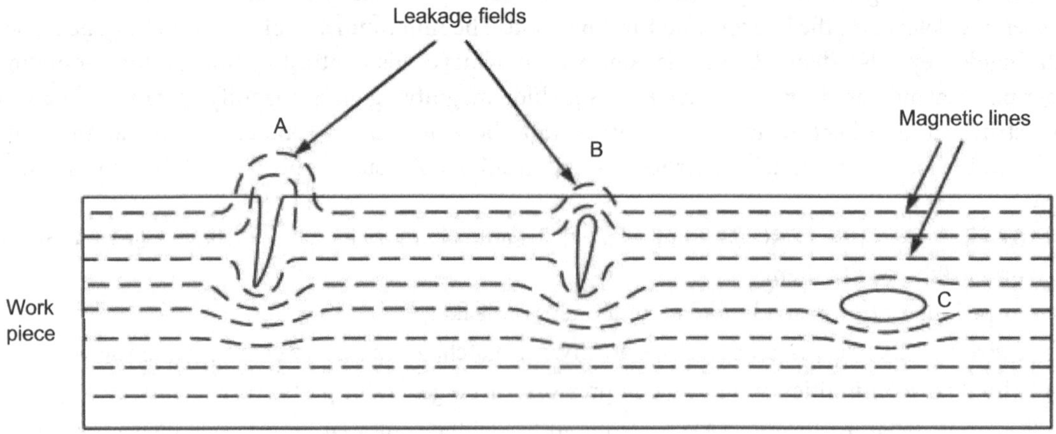

Fig. 4.14: Localised leakage of magnetic flux at flaws A and B.

As a consequence, magnetic particles pile up and bridge the discontinuities exist in the surface of the component. The magnetic bridge so formed delineates the discontinuity and generally indicates its location, size, shape and extent. If the magnetic bridge is fuzzy, it is an indication of a subsurface discontinuity.

Magnetic particles can be applied as dry powder or more commonly as particles suspended in a liquid such as kerosene. This liquid suspension is called magnetic ink. A discontinuity in the component is detectable if it is oriented nearly 90° with respect to the direction of the magnetic flux. For this the component must be rotated within the magnetic field such that the discontinuity becomes at 90° with respect to the direction of magnetic flux. In this condition the magnetic poles formed due to discontinuity are so close that they can attract the magnetic particles effectively to bridge the surface opening. In case the orientation is such that the magnetic poles are far of, i.e. when the surface crack is oriented parallel with respect to the magnetic lines of force (at location *C* in Figure 4.13), they cannot attract the magnetic particles effectively and the indication cannot be obtained. However, it is generally possible to detect the surface discontinuity even if it is oriented at 45° with respect the magnetic flux. The colour of the magnetic particles should be in good contrast with respect to the colour of the surface of the component.

MPI is one of the oldest and cheapest NDT-techniques which is extensively used to detect surface and shallow subsurface defects in the casting, forging, and welding industries of ferromagnetic materials such as iron, nickel, cobalt, and their suitable alloys. Recently, automation in MPI is also reported.

4.4.3 Liquid Penetrant Inspection

Liquid penetrant examination is simple, less expensive, fast and reliable, but its use is limited to those defects which are open to the surface of the component and is capable of giving only approximate depth and size of flaws. This technique is applicable to almost any component, whether it is small or large, of simple or complex shape. This technique can be carried out at any stage from in-process to final or during maintenance of a component. Liquid penetrant method is very reliable in detection of fatigue cracks which occur during service period of a material.

This method utilises the natural tendency of a fluid to segregate in its vicinity to a surface discontinuity and gives an indication of a crack or any other surface discontinuity. In case a surface discontinuity exists, liquid has the tendency to seep into it as compared to the surrounding area by capillary action. To detect a minute discontinuity, a good contrast must be developed between the background and the area of excess fluid. The basic principle of liquid penetrant inspection depends on the ability of the liquid penetrant to wet the surface of the component under test and flow over the surface to form a continuous and reasonably uniform coating, thus penetrating the surface discontinuities if any exist. The ability of a fluid to spread over a surface and penetrate the surface depends on certain factors. These factors include surface tension, viscosity, and capillary action of the fluid. Viscosity of a fluid affects the flowability of the fluid. Viscous fluids are difficult to spread over the surface and take more time to penetrate the surface discontinuities. Such discontinuities can be made visible by subsequent development. In order to achieve good defect visibility, the penetrating liquid will either be coloured with a bright and persistent dye or else contain a fluorescent compound. In the former type, the dye is generally red in colour and the developed surface can be viewed in natural or artificial light. In the latter case, on the other hand, the component must be viewed under ultraviolet light in order to visualise the defect. When the particles in the fluorescent fluid are struck by ultraviolet light, they are excited to a higher energy state. After being excited, these particles have natural tendency to come to the ground or the original energy state. In doing so they emit light radiations having wavelength longer than the original source. Thus, the emitted light is in the visible spectrum. Since the ultraviolet light is not visible to the human eye, it is also called black light.

4.4.4 Ultrasonic Testing

In ultrasonic testing, sound waves of high frequency are introduced into the material under inspection to detect subsurface and surface flaws. The sound waves travel through the material with some attendant loss of energy and are reflected back when they encounter an interface. The interface can be an airgap within the material or the opposite face of the material. By determining the relative times for the ultrasonic waves to be reflected back from the flaw and from the opposite face, the location of the flaw and its dimensions can be analysed. Flaws such as cracks, laminations, shrinkage cavities, pores, flakes and other discontinuities that act as metal-gas interfaces can easily be detected using ultrasonic waves. Inclusions and other inhomogeneities which cause partial reflections of sound waves can also be detected using ultrasonic waves. Ultrasonic inspections are carried out to detect both surface and internal flaws in the material. These techniques are used for the quality control inspection of semi-finished parts (such as blooms, billets, rolled slabs) as well as finished products. The technique is also in regular use for the in-service characterization of parts and assemblies.

The basic principle of ultrasonic testing depends on the transmission of energy of the ultrasonic waves to the particles of the medium through which they propagate and cause the displacement of particles. Usually, a transducer is used to generate the ultrasonic waves. A transducer consists of a piezoelectric crystal such as quartz which is subjected to an electric field. A variation in electric field will result in changes in dimensions of the crystal, thereby, producing both compression and tension in the crystal. The transducer is coupled to the test piece through a suitable liquid called couplant which enhances the sound energy transmission between the two. Either a continuous wave or pulses of ultrasonic frequencies are imposed on the transducer which in turn transmits the energy to the test piece. As the sound waves travel through the material, they lose some of the energy due to scattering at microscopic interfaces (such as grain boundaries, second phase particles, inclusions, etc.) and some due to internal friction of the material and are reflected at the interfaces. The reflected wave is detected by the original transducer or a second transducer and analysed to define the presence and location of flaws.

Among the various methods of ultrasonic inspection, following are the two commonly used methods:

(a) Normal incident pulse-echo method, and

(b) Normal incident through transmission method

The principle of pulse-echo method is that when a beam of ultrasonic wave is allowed to pass through a test piece, the echo is received either from a flaw or from the opposite surface of the test piece. The echo in the form of a pulse is converted to electrical energy and displayed on the screen of cathode ray oscilloscope as a pip (marker). In this method only one crystal probe is used for both sending and receiving the ultrasonic waves (Figure 4.14(a)).

In this transmission method, instead of a single transducer, two transducers are in use (Figure 4.15). One is placed in contact to the object through an oil film on the front or top surface to transmit the waves. The other is placed on the reverse side of the test piece through a couplant to receive the transmitted beam after it has travelled through the material. An electrical impulse of desired frequency is imposed on the transmitting crystal which produces a beam of ultrasonic wave.

This wave travels through the material and arrives to the reverse surface of the test piece where the receiver transducer receives it and converts it into an electrical signal. The electrical signal is amplified and displayed on the screen of cathode ray oscilloscope or indicated on another suitable indicating device.

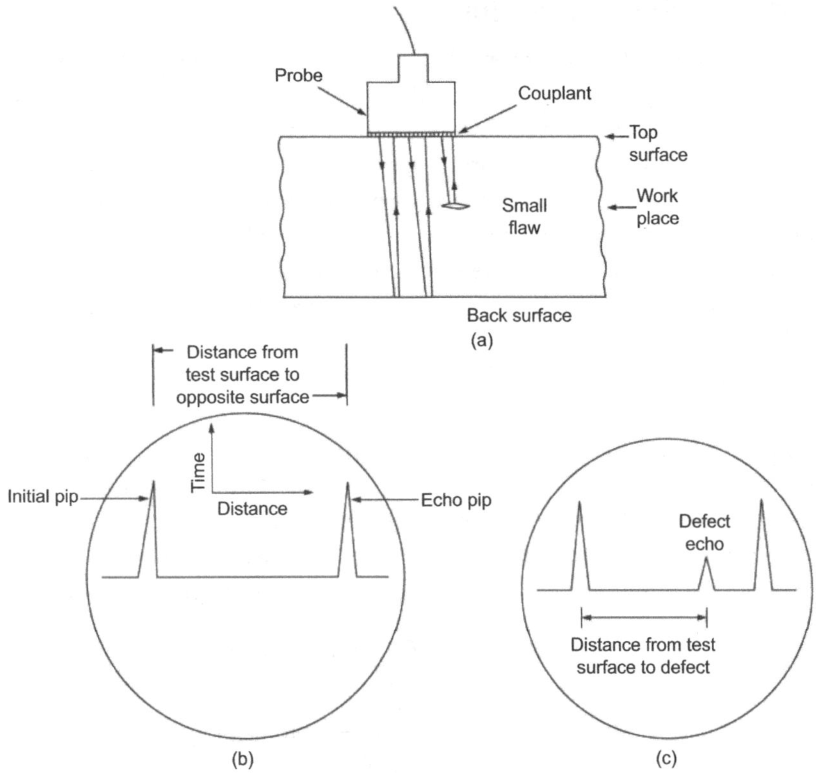

Fig. 4.15: Illustration of pulse-echo method. (a) the probe acts both as transmitter and receiver, (b) display of reflection of ultrasonic waves from the top and back surface, (c) intermediate pip showing presence of defect in the workpiece.

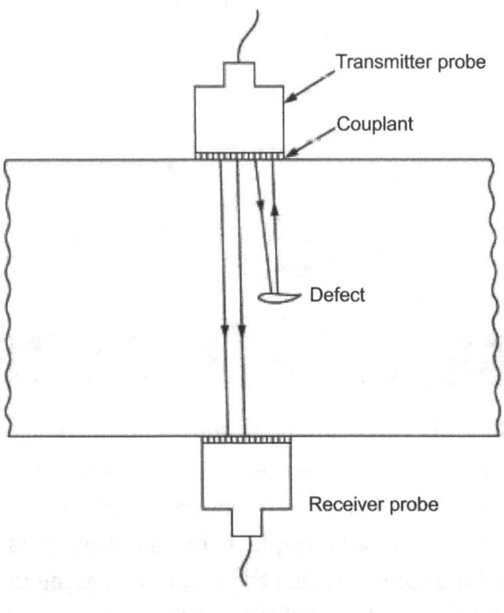

Fig. 4.16: Illustration of normal probe transmission method of ultrasonic testing

4.4.5 Radiographic Inspection

Radiographic inspection makes use of very short wavelength electromagnetic radiations, namely, X-rays and γ-rays. These rays are generally penetrable through the solid materials and are partially absorbed by the medium (Fig. 4.16). The amount of absorption that occurs will depend on the density and thickness of the material through which the radiation passes and also on the characteristic of the radiation. Consequently, there will be a variation in the absorption of the rays by the material in the defective area. The radiation that passes through the material can be detected and recorded on either a photographic film or viewed on a fluorescent screen or detected and monitored by electronic sensing equipment. Thus, the term radiography can be defined as a process in which an image is produced on a radiation sensitive film. This technique is one of the most widely used NDT methods for the detection of internal defects such as porosity, inclusion, and voids. With proper orientation planar defects can also be detected with radiography.

The basic principle of radiographic inspection is that the object to be examined is placed in the path of the beam of radiation of short wavelength (X-ray or g-ray) source. A photographic film is placed on the other side of the object (from the opposite side of the beam source). Some of the radiations will be absorbed by the material, but some will travel through the object and impinge on the film, producing a latent image. When the film is developed there will be a picture of light and dark areas.

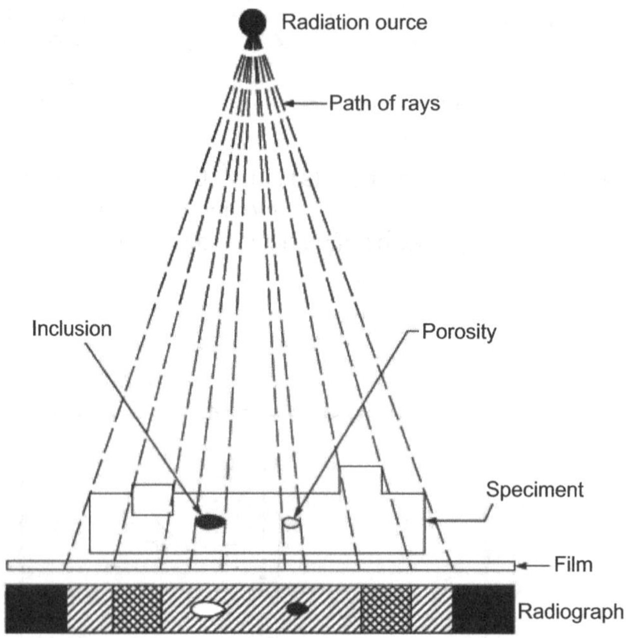

Fig. 4.17: Radiograph showing differential absorption.

The dark areas represent the parts of the object having lower density. These parts may be the flaws such as pore or void in the material. This film in general is called radiograph (specifically a xerograph if the film is exposed to X-rays or a radiograph if the radiation used is g (gamma)-ray). Figure 4.16 illustrates the principle of radiography in which the specimen having the stepped shape block with thick and thin steps and containing pore and an inclusion is exposed to short wavelength radiations.

The developed radiograph shows regions of variation in darkness. The thin section of the object (specimen) allowed more radiations to pass through, and therefore, appear dark region on the radiograph. The same effect is found in the area of porosity, but this region is darkest in the radiograph. On the other hand, the inclusion, being denser than the surrounding material, absorbs more radiations and therefore produces light area on the photographic film. Basically, it is this contrast (different in density) on the film that allows an observer to detect flaws in the specimen.

The industrial radiography is commonly employed to examine the welded products (such as weld in pressure vessels, bridge construction, power projects, etc.) and castings for blow holes and other flaws. Radiography can also be used to measure the thickness of the material.

UNIT SUMMARY

1. The present chapter discusses the problem-solving models to be adopted for failure analysis and different steps for determination of root cause of failure and preventive measures.
2. The different facture modes due to brittle and ductile fracture are explained.
3. The concept of fatigue failure, significance of S-N curve and factors affecting the fatigue strength of the material are also discussed.
4. The failure due to creep and different regions present in the creep curve is also briefly explained.
5. The basic difference between destructive and non-destructive tests have been explained.
6. To have more practical approach, the methods and machines used to perform these tests have been explained in detail.

EXERCISES

Multiple Choice Questions

1. Failure analysis techniques can include:
   (a) Visual inspection (b) Statistical analysis
   (c) Non-destructive testing (d) All
2. Ductile fracture is characterized by:
   (a) Rapid, brittle fracture (b) Slow, plastic deformation
   (c) Shear failure along a cleavage plane (d) Low-energy impact resistance
3. The primary mechanism of ductile fracture is:
   (a) Cleavage (b) Intergranular
   (c) Transgranular (d) Fatigue failure
4. Ductile fracture is typically characterized by:
   (a) Smooth, flat fracture surfaces
   (b) shining fractured surfaces
   (c) Irregular fracture surfaces with voids or dimples
   (d) No visible fracture surfaces

5. The fracture surface in the brittle fracture is characterized by:
   (a) Smooth and flat surfaces (b) Dimples
   (c) shining fractured surfaces (d) All
6. The tendency of brittle fracture increases with:
   (a) Decreasing temperature (b) Increasing temperature
   (c) Decrease in strain rate. (d) It doesn't depend on temperature or strain rate.
7. Which of the following show a fatigue limit?
   (a) Titanium (b) Cast iron
   (c) Magnesium (d) Al-Mg alloys
8. What cycles range is chosen for the endurance limit for mild steel?
   (a) $10^2 - 10^3$ (b) $10^5 - 10^6$
   (c) $10^7 - 10^8$ (d) $10^{11} - 10^{12}$
9. What decreases fatigue strength?
   (a) Shot peening (b) Case hardening
   (c) Inclusions (d) cold working
10. When fatigue occurs above __ cycles, it is called high-cycle fatigue.
    (a) 10 (b) 10^3
    (c) 10^2 (d) 10^8
11. In which stage of creep, the creep rate is constant?
    (a) Primary creep (b) Secondary creep
    (c) Tertiary creep (d) None
12. In which stage of creep, the creep rate is maximum?
    (a) Primary creep (b) Secondary creep
    (c) Tertiary creep (d) None
13. Creep depends on
    (a) Temperature (b) Time
    (c) Tensile Load (d) All of these
14. In tensile testing of ductile material, necking starts at
    (a) Lower Y.S. (b) Upper Y.S.
    (c) UTS (d) Just before the fracture
15. Which of the following is a destructive testing methods
    (a) Metallographic testing (b) Hardness testing
    (c) Tensile testing (d) Radiography
16. Ultrasonic testing is done on materials to determine
    (a) Cracks below the surface (b) Y.S.
    (c) U.T.S. (d) Hardness
17. The natural force which makes Liquid Penetrant Test possible is known as
    (a) Charle's Law (b) Gravitational Law
    (c) Surface tension (d) Capillary action

18. Which of the following is limitation of the liquid penetrant testing
- (a) It can't detect the sub-surface discontinuities
- (b) It can't be used on nonmagnetic materials
- (c) It can't be used on the ferromagnetic materials
- (d) All of the above

19. Materials that are commonly inspected using Liquid Penetrant Testing
- (a) Plastics
- (b) Metals
- (c) Ceramics
- (d) All of the above

20. What are the factors that influence the Fatigue life of materials?
- (a) Nature and magnitude of the stress cycle
- (b) Endurance limit
- (c) Surface characteristics
- (d) All of the above

21. To examine areas inside a pipe section, the visual examiner uses a
- (a) telescope
- (b) torch
- (c) borescope
- (d) microscope

22. Which NDT method is mostly used to check the sub-surface lamination in a rolled plates
- (a) Radiography test
- (b) Ultrasonic test
- (c) Liquid penetrant test
- (d) All

23. Which type of non-destructive test is fit to check the inclusion present in rolled steel bar?
- (a) Visual test
- (b) Liquid penetrant test
- (c) Radiography test
- (d) All

24. Which of the following steps are involved in intelligent visual inspection using AI tool
- (a) Training
- (b) Validation
- (c) Testing
- (d) All

ANSWERS															
1.	(d)	2.	(b)	3.	(c)	4.	(c)	5.	(a)	6.	(a)	7.	(b)	8.	(c)
9.	(c)	10.	(b)	11.	(b)	12.	(c)	13.	(d)	14.	(c)	15.	(a)	16.	(a)
17.	(d)	18.	(a)	19.	(d)	20.	(d)	21.	(c)	22.	(b)	23.	(c)	24.	(d)

Short and Long Answer Type Questions

1. Differentiate between ductile and brittle fracture of a metal. Use clear schematic to indicate the characteristic features of both types of fracture.
2. What do you understand by failure analysis? Discuss the commonly used steps involved in the problem-solving model for root cause failure analysis.
3. Discuss the mechanism of brittle fracture for brittle materials. What are the parameters which will affect brittle fracture of a material?
4. Why Orowan and Irwin modification is suggested to understand the brittle fracture?
5. What do you understand by fatigue fracture? Discuss the factors affecting fatigue property of metallic materials.
6. What do you understand by endurance limit in fatigue testing? How do you determine this value for a given metal?

7. Why does S-N curve commonly used to understand fatigue behaviour of a metal? Draw S-N curves for mild steel and high strength aluminium alloy.
8. Where do fatigue failures commonly originate on a metal section?
9. What is creep? Use clear schematic to describe the creep behaviour of metals.
10. What is difference between creep testing and tensile testing?
11. How creep is different from fatigue failure?
12. Compare and contrast between destructive testing and non-destructive testing.
13. Distinguish between elastic and plastic deformation.
14. Define engineering stress. State the SI unit of engineering stress and engineering strain.
15. What is the relation between engineering strain and percentage elongation?
16. Define the hardness of metal.
17. What type of indenters are used in (a) Brinell hardness test, (b) Rockwell B hardness, and (c) Rockwell C hardness?
18. What will be the effect of using 5mm and 10mm hardened steel ball in case of performing Brinell hardness testing of high strength steel plate?
19. Suppose you have to check the hardness of a cold rolled mild steel sheet of thickness 5mm. Which hardness testing method will be utilized by you and why?
20. State the application of visual inspection. How artificial intelligence (AI) is being utilized to improve the efficiency and effectiveness of visual inspection?
21. Describe the working principle of magnetic particle inspection (MPI). Use clear schematic to describe MPI.
22. Describe the working principle of liquid penetrant inspection (LPI). Use clear schematic to describe LPI.
23. Suppose you have to track the surface crack of mild steel sample and high strength aluminium alloy. Which NDT method will be utilized for this purpose and why?
24. Describe the working principle of ultrasonic testing. Use clear schematic to describe commonly used ultrasonic testing methods. Write down the application of ultrasonic testing methods.
25. Describe the working principle of radiographic inspection. Use clear schematic to describe this method and write down the industrial application of radiographic inspection.
26. Suppose we have to test the presence of inclusion in a mild steel. Which NDT method will be preferred to perform this assignment and why?

PRACTICAL

Experiment title: To perform the tensile test of mild steel.

Objectives:

(a) Plot the engineering stress vs engineering strain curve for the mild steel using tensile data.

(b) Determine yield strength, tensile strength, percent elongation and percent reduction in area from the tensile plot.

Apparatus/Instrument: Tensile testing machine, mild steel sample.

Theory:

Students can refer to ASTM E8M standard, which describes the tensile testing of metals at room temperature.

Procedure:

Students can refer to ASTM E8M for standard guide for sample preparation for tensile testing of metals.

Result:

Material:

Original dimensions: Length .... Diameter ....
Final dimensions: Length .... Diameter.......................
The students should plot the engineering stress vs engineering strain curve from the tensile data in the space provided below.

Load (Kg)	Displacement (mm)	Engineering stress (Kg/mm2)	Engineering strain

Conclusion:

The values of the yield strength, tensile strength, percent elongation and percent reduction in area for mild steel are ....

KNOW MORE

Students can refer to advanced testing techniques such as temperature dependent tensile test, and high strain rate test.

REFERENCES AND SUGGESTED READINGS

1. R. Balasubramaniam; Callister's Materials Science and Engineering, Wiley India Pvt. Ltd., 2nd edition 2014.
2. V. Raghavan; Materials Science and Engineering- A first course, Prentice Hall India Learning Private Limited, 6th edition 2015.
3. George E. Dieter, Mechanical Metallurgy, McGraw Hill Education, 3rd edition, 2017.
4. A.K. Bhargava, C.P. Sharma, Mechanical Behavior and Testing of Materials, PHI Learning, 1st edition, 2011.

Dynamic QR code for further reading

5 Corrosion and Surface Engineering

UNIT SPECIFICS

Through this unit the following aspects of corrosion and surface engineering has been discussed:

- Nature of corrosion and factors affecting corrosion;
- Different forms of corrosion and their mechanism;
- Corrosion control: materials selection, design, coatings, inhibitors, cathodic protection;
- Significance of surface engineering and surface treatments;
- Organic and metallic coatings.

The fundamental aspect of corrosion engineering and various protection mechanisms has been incorporated in the text to impart its importance to the students. The concept and impact of surface engineering and different methods of surface treatment and protection are also provided which provide their practical significance to the students.

Besides giving many multiple-choice questions as well as questions of short and long answer types marked in two categories following lower and higher order of Bloom's taxonomy, a list of references and suggested readings are given in the unit so that one can go through them for further advanced knowledge.

This section also highlights the examples of some interesting facts, applications of the subject matter for industrial applications on variety of aspects and/or our day-to-day real life, case study related to environmental, sustainability, social and ethical issues (whichever applicable), and finally inquisitiveness associated with the topics of the unit.

RATIONALE

This unit helps students to have a fundamental understanding of corrosion and methods to mitigate them. The different forms of corrosion and mechanisms are well explained with the help of schematics and figures. The different protection methods based on materials selection, design, inhibitors, coatings and cathodic protection have been explained in great depth. It then explains the significance of surface engineering and surface treatments for enhanced properties of the engineering components. The different deposition techniques such as electroplating and electroless plating and methods of deposition such as PVD and CVD also have been discussed. The unit helps the students to explore the understanding of material-environment interaction and selection of suitable materials/treatment to be used in a given specific environment. Finally, student would be able to correlate and contribute the knowledge of corrosion control and surface engineering in the relevant industries.

PRE-REQUISITES

Basic electrochemistry.

UNIT OUTCOMES

List of outcomes of this unit is as follows:

U5-O1: Explain different forms of corrosion

U5-O2: Acquire knowledge about various corrosion control methods

U5-O3: Have understanding about different surface treatments and coatings

U5-O4: Know about the various coating deposition techniques and methods

Unit-5 Outcomes	EXPECTED MAPPING WITH COURSE OUTCOMES (1- Weak Correlation; 2- Medium correlation; 3- Strong Correlation)				
	CO-1	CO-2	CO-3	CO-4	CO-5
U5-O1	3	2	2	1	3
U5-O2	3	2	2	1	3
U5-O3	3	2	2	1	3
U5-O4	3	2	2	1	3

5.1 NATURE OF CORROSION AND ITS CAUSES

Corrosion is not merely an engineering issue, but it also leads to global economic loss. In a recent report of National Association of Corrosion Engineers (NACE) International, the global economic loss due to corrosion has been evaluated to \$2.5 trillion, which covers ~3% of the global gross domestic product (GDP). Among the various engineering sectors, petroleum, oil and gas and construction industries are the major contributors to the corrosion cost globally. It is, therefore, essential for an engineer to understand the basic principles of corrosion and the methods of protection against corrosion.

Corrosion is defined as the destruction or deterioration of a material because of the reaction it undergoes with its environment. corrosion predominantly occurs when a material prone to corrosion comes in contact with a corrodent. For example, when mild steel (contains iron) comes in contact with the oxygen and moisture present in the air forms rust, which is nothing but corrosion. For the corrosion to take place, the formation of a corrosion cell is essential, which comprises of four different components namely an anode, cathode, electrolyte and metallic path. Figure 5.1 shows the schematic of the corrosion cell formation. The anode in the corrosion cell undergoes oxidation and liberates electrons. These electrons are consumed at the cathode and leads to reduction. The electrolyte is an electrically conductive solution that must be present for the corrosion to occur. Both the anode and the cathode are connected externally by a metallic conductor which provides the path for the flow of electric current.

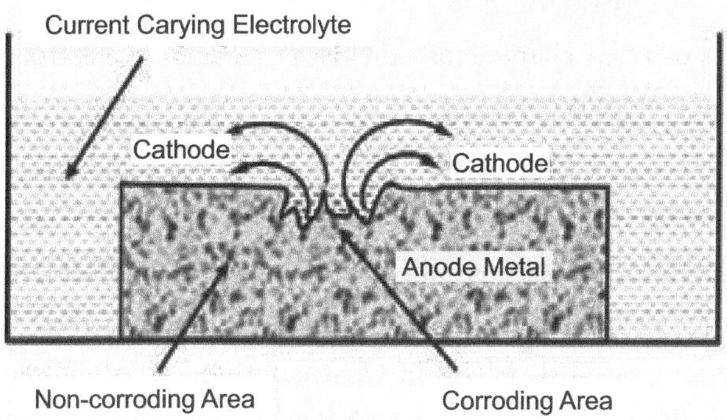

Fig. 5.1: Schematic of a corrosion cell.

Corrosion can be classified as (a) wet corrosion and (b) dry corrosion. Wet corrosion occurs in the presence of liquids such as aqueous solutions and electrolytes. An example is corrosion of steel in water. Dry corrosion occurs above the dew point of the environment in the absence of a liquid phase. The major corrodents are vapors and gases and which are associated with high temperature. The common example is corrosion of steel by furnace gases.

Corrosion environments include the atmosphere, aqueous solutions, soils, acids, bases, inorganic solvents, molten salts, liquid metals, and fluids present in human body. On a tonnage basis, atmospheric corrosion accounts for the greatest losses. Moisture containing dissolved oxygen is the primary corrosive agent, but other substances, including sulfur compounds and sodium chloride, may also contribute. This is especially true for marine atmospheres, which are highly corrosive because of the presence of sodium chloride. Dilute sulfuric acid solutions (acid rain) in industrial environments can also cause corrosion problems. Metals commonly used for atmospheric applications include alloys of aluminium and copper, and galvanized steel.

Water environments can also have a variety of compositions and corrosion characteristics. Freshwater normally contains dissolved oxygen as well as minerals, several of which account for hardness. Seawater contains approximately 3.5% salt (predominantly sodium chloride), as well as some minerals and organic matter. Seawater is generally more corrosive than freshwater, frequently producing pitting and crevice corrosion. Cast iron, steel, aluminium, copper, brass, and some stainless steels are generally suitable for freshwater use, whereas titanium, brass, some bronzes, copper–nickel alloys, and nickel-chromium-molybdenum alloys are highly corrosion resistant in seawater.

Soils have a wide range of compositions and susceptibilities to corrosion. Compositional variables include moisture, oxygen, salt content, alkalinity, and acidity, as well as the presence of various forms of bacteria. Cast iron and plain carbon steels, both with and without protective surface coatings, are economical for underground structures.

5.2 ELECTROCHEMICAL REACTIONS

For metallic materials, the corrosion process is normally electrochemical in which there is transfer of electrons from one chemical species to another. Metal atoms characteristically lose or give up electrons in what is called an oxidation reaction. For example, a metal M that has a valence of *n* could experience oxidation according to the reaction

$$M \rightarrow M^{n+} + ne^- \quad (5.1)$$

where M becomes an n positively charged ion and loses its n valence electrons; e is used to symbolize an electron. Examples in which metals oxidize are

$$Fe \rightarrow Fe^{2+} + 2e^- \quad (5.2)$$

$$Al \rightarrow Al^{3+} + 3e^- \quad (5.3)$$

The site at which oxidation takes place is called the anode oxidation which is sometimes called an anodic reaction. The electrons generated from each metal atom that is oxidized must be transferred to and become a part of another chemical species in what is termed a reduction reaction. For example, some metals undergo corrosion in acid solutions, which have a high concentration of hydrogen (H) ions; the H ions are reduced as follows:

$$2H^+ + 2e^- \rightarrow H_2 \quad (5.4)$$

and hydrogen gas (H_2) is evolved.

Other reduction reactions are possible, depending on the nature of the solution to which the metal is exposed. For an acid solution having dissolved oxygen, reduction according to the following reaction will occur.

$$O_2 + 4H^+ + 4e^- \rightarrow 2H_2O \quad (5.5)$$

For a neutral or basic aqueous solution in which oxygen is also dissolved,

$$O_2 + 2H_2O + 4e^- \rightarrow 4(OH^-) \quad (5.6)$$

An overall electrochemical reaction must consist of at least one oxidation and one reduction reaction and will be their sum; often the individual oxidation and reduction reactions are termed half-reactions. There can be no net electrical charge accumulation from the electrons and ions and the total rate of oxidation must equal the total rate of reduction, or all electrons generated through oxidation must be consumed by reduction. Cathodic reaction to be introduced

For example, consider zinc metal immersed in an acid solution containing H+ ions.

At some regions on the metal surface, zinc will experience oxidation or corrosion as illustrated in Figure 5.2, according to the reaction

$$Zn \rightarrow Zn^{2+} + 2e^- \quad (5.7)$$

The corresponding reduction reaction is the hydrogen evolution reaction which can be written as

$$2H^+ + 2e^- \rightarrow H_2 \quad (5.8)$$

If no other oxidation or reduction reactions occur, the total electrochemical reaction can be written as

$$Zn + 2H^+ \rightarrow Zn^{2+} + H_2 \quad (5.9)$$

Another example is the oxidation or rusting of iron in water, which contains dissolved oxygen. This process occurs in two steps; in the first, Fe is oxidized to Fe^{2+} [as $Fe(OH)_2$],

$$Fe + \tfrac{1}{2} O_2 + H_2O \rightarrow Fe^{2+} + 2OH^- \rightarrow Fe(OH)_2 \quad (5.10)$$

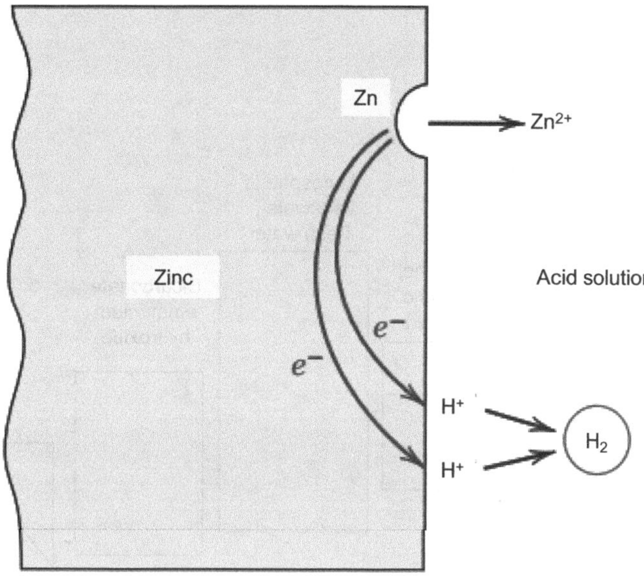

Fig. 5.2: The electrochemical reactions associated with the corrosion of zinc in an acid solution.

and in the second stage to Fe^{3+} [as $Fe(OH)_3$] according to the reaction

$$2\ Fe(OH)_2 + \frac{1}{2}\ O_2 + H_2O \rightarrow 2\ Fe(OH)_3 \tag{5.11}$$

As a consequence of oxidation, the metal ions may either go into the corroding solution as ions ($Fe(OH)_2$) or form an insoluble compound with nonmetallic elements ($Fe(OH)_3$).

5.3 FACTORS AFFECTING CORROSION

The following are the factors which affect the corrosion of a material;

5.3.1 Environment

5.3.1.1 Oxidizing/reducing characteristics

Environments can be characterized by their oxidizing power. An environment that is strongly oxidizing has a great tendency to oxidize materials in contact with it, while an environment that is strongly reducing has a much lesser tendency to oxidize materials. Figure 5.3 illustrates the oxidizing/reducing power of several common environments. Concentrated nitric acid and wet chlorine environments are strongly oxidizing and promote the oxidation of almost all metals except the noble metals. The oxidizing power of sulfuric acid is a function of concentration and ranges from moderately reducing for dilute acids to more strongly oxidizing for concentrated acids. Deaerated hydrochloric acid is a strongly reducing environment, and the addition of oxidizing species such as dissolved oxygen or ferric (Fe^{3+}) ions would significantly increase oxidizing power.

Atmospheric, seawater, and freshwater environments are typically nearly neutral and range from moderately reducing to moderately oxidizing. Sodium hydroxide is a strongly alkaline environment represented by high pH, and bicarbonate and ammonium hydroxide solutions are weak bases with comparatively lower pH values.

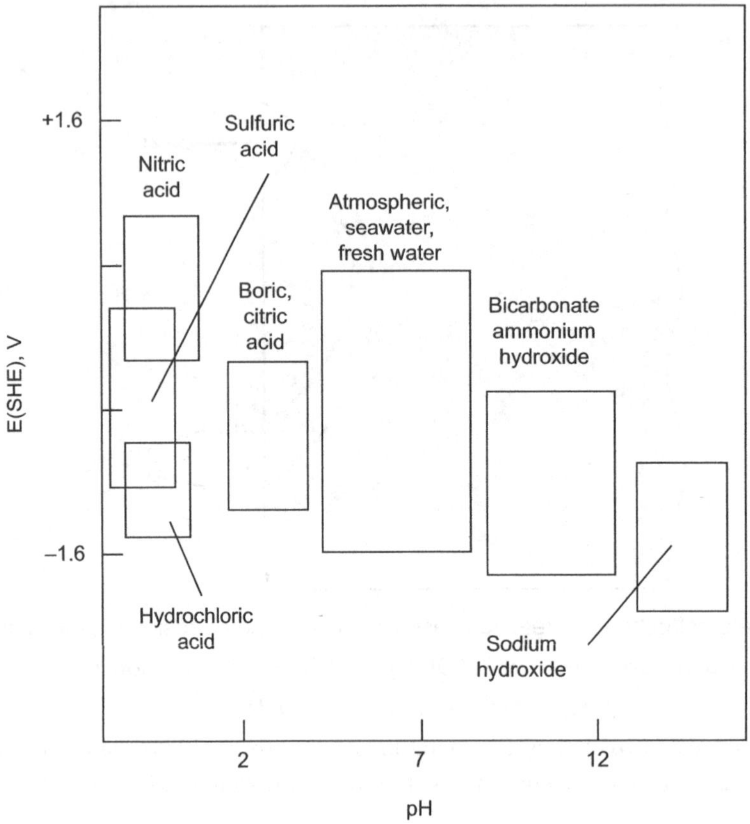

Fig. 5.3: Approximate locations of several common environments on a potential-pH diagram. SHE, standard hydrogen electrode

5.3.1.2 Temperature effect

Corrosion is an activation-controlled electrochemical reaction and the rate is greatly affected by temperature. Typically, corrosion rate increases significantly as temperature increases. A rule of thumb is that the corrosion rate will double for each 10°C (18 °F) increase in temperature. The increased corrosion rate results from increased activation energy for chemical and electrochemical reactions, increased diffusion rates in the electrolyte, and increased transport through the electrolyte or environment and across films that could have formed on the metal surface. There are observed reversals in the trends regarding temperature and corrosion rate. These are typically related to the stability of passive films that may form or break down over a fairly narrow temperature range and also could be related to the solubility of corrosive or beneficial gases and species in the environment.

An important consideration with respect to temperature is the temperature of the metal surface in contact with the environment. Often the surface temperature of a heating element or heat-exchanger component can be significantly greater than the bulk environment temperature. This can result in significantly increased corrosion rates of these surfaces with respect to the corrosion rate of unheated metal surfaces in contact with the same electrolyte.

5.3.1.3 Presence of detrimental species

The presence of individual species in an environment, even at trace levels, could have a major impact on the corrosion behavior of the given material. Chlorides are particularly detrimental to materials that depend on a passive film for their corrosion resistance (e.g., stainless steels). In the presence of chloride ion, the passive film can break down locally and result in pitting, concentration-cell formation, and in some cases stress-corrosion cracking (SCC). The presence of ammonia in an environment can greatly increase the corrosion of some alloys. The ammonia ion forms soluble complexes with copper in the metal and results in the breakdown of passive surface films on copper alloys and more rapid corrosion.

5.3.1.4 Effect of velocity

The influence of fluid flow rate, or fluid velocity, is a complex variable, and its influence on corrosion behavior is dependent on the alloy, fluid constituents, fluid physical properties, geometry, and corrosion mechanism. The presence of fluid flow can sometimes be beneficial in preventing or decreasing localized attack. For example, type 316 stainless steel has been shown to pit in stable seawater but not in moving seawater. When the seawater is moving, the mass transfer rate of oxygen is high enough to maintain a completely passive surface, but in the absence of flow, the mass transfer of oxygen is too slow and the surface cannot remain passive. Under other circumstances, fluid flow can cause a type of erosion of a surface through the mechanical force of the fluid itself. This common process is called impingement, and an example of such erosion is where fluid is forced to change its flow direction at pipe bends. When solids are present in the liquid, they can cause erosion-corrosion.

5.3.2 Metallurgical Factor

Cold working or plastic deformation is used in ductile metals to increase their strength; however, a cold-worked metal is more susceptible to corrosion than the same material in an annealed state. For example, deformation processes are used to shape the head and point of a nail; consequently, these positions are anodic with respect to the shank region. Thus, differential cold working on a structure should be a consideration when a corrosive environment may be encountered during service.

5.4 FORMS OF CORROSION

Metallic corrosion could be classified into eight forms: uniform, galvanic, crevice, pitting, intergranular, selective leaching, erosion–corrosion, and stress corrosion. The causes and means of prevention of each of these forms are discussed briefly.

5.4.1 Uniform attack

Uniform attack is a form of electrochemical corrosion that occurs uniformly over the entire exposed surface and often leaves behind a scale or deposit. In a microscopic sense, the oxidation and reduction reactions occur randomly over the surface. Few of the examples include general rusting of steel and iron and the tarnishing of silverware. This is probably the most common form of corrosion. It is also the least objectionable because it can be predicted and designed with relative ease. Figure 5.4 illustrates the uniform attack due to rusting of metal roof. Uniform attack can be prevented by (a) proper selection of materials including coatings, (b) inhibitors or (c) cathodic protection.

Fig. 5.4: Uniform corrosion of metal roof

5.4.2 Galvanic corrosion

Galvanic corrosion occurs when two metals or alloys having different compositions are electrically coupled while exposed to an electrolyte. A potential difference usually exists between two dissimilar metals when they are immersed in a corrosive or conductive solution. If these metals are placed in contact, this potential difference produces electron flow between them. Corrosion of the less corrosion resistant metal is usually increased and attack of the more resistant material is decreased, as compared with the behavior of these metals when they are not in contact. The less resistant metal becomes anodic and the more resistant metal cathodic. Usually, the cathode or cathodic metal undergoes least dissolution during the coupling. Because of the electric currents and dissimilar metals involved, this form of corrosion is called galvanic, or two-metal corrosion.

The rate of galvanic attack depends on the relative anode-to-cathode surface areas that are exposed to the electrolyte, and the rate is related directly to the cathode- anode area ratio—that is, for a given cathode area, a smaller anode corrodes more rapidly than a larger one because corrosion rate depends on current density. Figure 5.5 represents the galvanic corrosion occurred in case of an oil and gas pipeline.

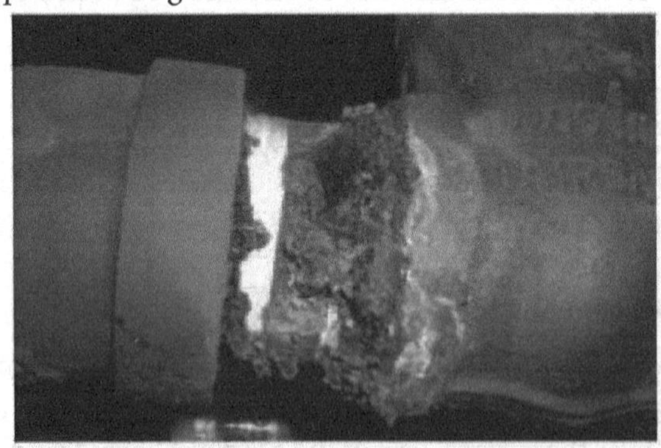

Fig. 5.5: An example of galvanic corrosion in oil and gas pipelines

Several measures may be taken to reduce the effects of galvanic corrosion significantly including the following:

1. If coupling of dissimilar metals is necessary, choose two metals that are close together in the galvanic series.

2. Avoid an unfavourable anode-to-cathode surface area ratio; use an anode area as large as possible.
3. Electrically insulate dissimilar metals from each other.
4. Electrically connect a third, anodic metal to the other two; this is a form of cathodic protection.

5.4.3 Crevice corrosion

The corrosion may also occur as a consequence of concentration differences of ions in the electrolyte solution and between two regions of the same metal piece. For such a concentration cell, corrosion occurs in the region that has the lower concentration. A good example of this type of corrosion occurs in crevices and recesses or under deposits of dirt or corrosion products where the solution becomes stagnant and there is localized depletion of dissolved oxygen. Corrosion preferentially occurring at these positions is called crevice corrosion. The crevice must be wide enough for the solution to penetrate yet narrow enough for stagnancy; usually the width is several thousandths of an inch. Figure 5.6 shows the presence of crevice in a gas pipeline.

Fig. 5.6: Oil and gas pipeline under crevice corrosion

The proposed mechanism for crevice corrosion is illustrated in Figure 5.7. After oxygen has been depleted within the crevice, oxidation of the metal occurs at this position. Electrons from this electrochemical reaction are conducted through the metal to adjacent external regions, where they are consumed by reduction. In many aqueous environments, the solution within the crevice has been found to develop high concentrations of H^+ and Cl^- ions, which are especially corrosive. Many alloys that passivate are susceptible to crevice corrosion because protective films are often destroyed by the H^+ and Cl^- ions.

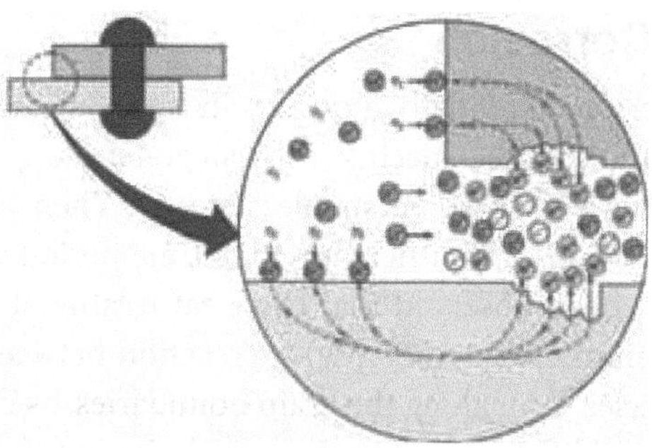

Fig. 5.7: Schematic illustration for mechanism of crevice corrosion between two riveted joints.

Crevice corrosion could be prevented by using welded instead of riveted or bolted joints, using non-absorbing gaskets when possible, removing accumulated deposits frequently, and designing containment vessels to avoid stagnant areas and ensure complete drainage.

5.4.4 Pitting Corrosion

Pitting is another form of very localized corrosion attack in which small pits or holes form. They ordinarily penetrate from the top of a horizontal surface downward in a nearly vertical direction. It is an extremely insidious type of corrosion, often going undetected and with very little material loss until failure occurs. An example of pitting corrosion is shown in Figure 5.8.

Fig. 5.8: Pitting corrosion on pipelines

The mechanism for pitting is similar to that of the crevice corrosion wherein oxidation occurs within the pit itself, and reduction at the surface. It is supposed that gravity causes the pits to grow downward, the solution at the pit tip becoming more concentrated and denser as pit growth progresses. A pit may be initiated by a localized surface defect such as a scratch or a slight variation in composition. In fact, it has been observed that specimens having polished surfaces display a greater resistance to pitting corrosion. Stainless steels are somewhat susceptible to this form of corrosion; however, alloying with about 2% molybdenum enhances their resistance significantly.

5.4.5 Intergranular Corrosion

Intergranular corrosion occurs preferentially along grain boundaries for some alloys and in specific environments. The net result is that the specimen disintegrates along its grain boundaries. This type of corrosion is frequently observed in some stainless steels. When heated to temperatures between 500°C and 800°C (950F and 1450F) for sufficiently long time periods, these alloys become sensitized to intergranular attack. It has been observed that this heat treatment permits the formation of small precipitate particles of chromium carbide ($Cr_{23}C_6$) by reaction between the chromium and carbon in the stainless steel. These particles form along the grain boundaries, as illustrated in Figure 5.9. Both the chromium and the carbon must diffuse to the grain boundaries to form the precipitates, which leaves a chromium depleted zone adjacent to the grain boundary. Consequently, this grain boundary region is now highly susceptible to corrosion.

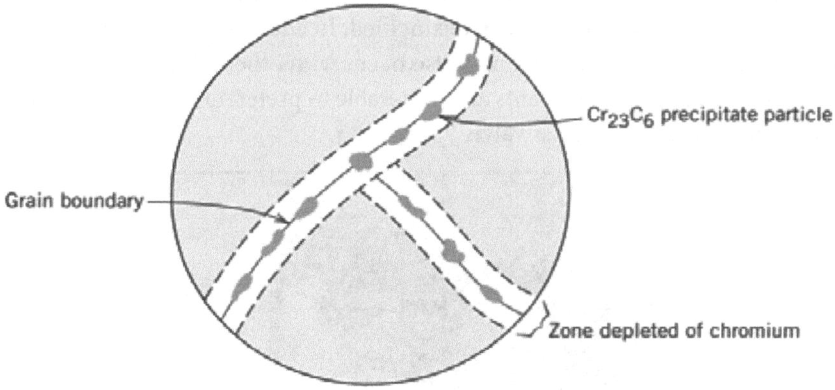

Fig. 5.9: Schematic illustration of chromium carbide particles that have precipitated along grain boundaries in stainless steel.

Intergranular corrosion is an especially severe problem in the welding of stainless steels, where it is often termed weld decay. Figure 5.10 shows this type of intergranular corrosion.

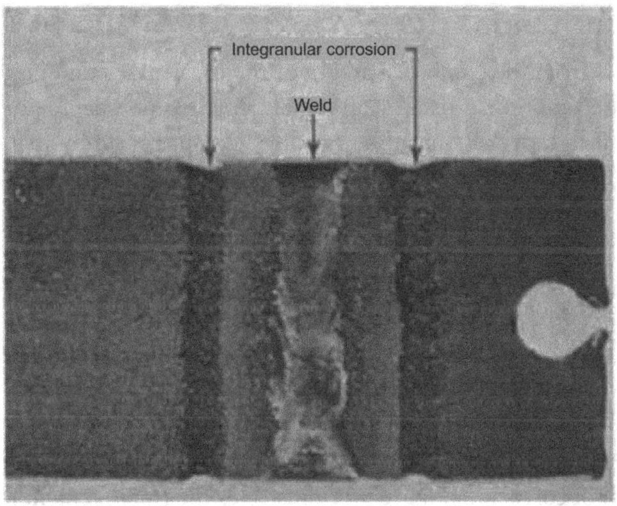

Fig. 5.10: Weld decay in a stainless steel. The regions along which the grooves have formed were sensitized as the weld cooled.

Stainless steels may be protected from intergranular corrosion by the following measures: (1) subjecting the sensitized material to a high-temperature heat treatment in which all the chromium carbide particles are redissolved, (2) lowering the carbon content below 0.03 wt% C so that carbide formation is minimal, and (3) alloying the stainless steel with another metals such as niobium or titanium, which has a greater tendency to form carbides than chromium. This process is known as stabilization.

5.4.6 Selective leaching

Selective leaching is found in solid solution alloys and occurs when one element or constituent is preferentially removed as a consequence of corrosion processes. The most common example is the dezincification of brass (shown in Fig. 5.11), in which zinc is selectively leached from a copper–zinc brass alloy. The mechanical properties of the alloy are significantly impaired because only a porous mass of copper remains in the region that has been dezincified. In addition, the material changes from yellow to a red or copper color. Selective leaching may also occur with other alloy systems in which aluminium, iron, cobalt, chromium, and other elements are vulnerable to preferential dissolution. Figure 5.11 shows the example of dezincification of a brass valve.

Fig. 5.11: Dezincification of a brass valve

5.4.7 Erosion corrosion

Erosion corrosion arises from the combined action of chemical attack and mechanical abrasion or wear as a consequence of fluid motion. Virtually all metal alloys, to one degree or another, are susceptible to erosion–corrosion. It is especially harmful to alloys that passivate by forming a protective surface film; the abrasive action may erode away the film, leaving exposed a bare metal surface. If the coating is not capable of continuously and rapidly re-forming as a protective barrier, corrosion may be severe. Relatively soft metals such as copper and lead are also sensitive to this form of attack. Usually, it can be identified by surface grooves and waves having contours that are characteristic of the flow of the fluid. The nature of the fluid could also have a significant effect on the corrosion behavior. Increasing the fluid velocity normally enhances the rate of corrosion. Also, a solution is more erosive when bubbles and suspended particulate solids are present. Erosion–corrosion is commonly found in piping, especially at bends, elbows, and abrupt changes in pipe diameter—positions where the fluid changes direction or flow suddenly becomes turbulent. Propellers, turbine blades, valves, and pumps are also susceptible to this form of corrosion. Figure 5.12 illustrates the impingement failure of an elbow fitting.

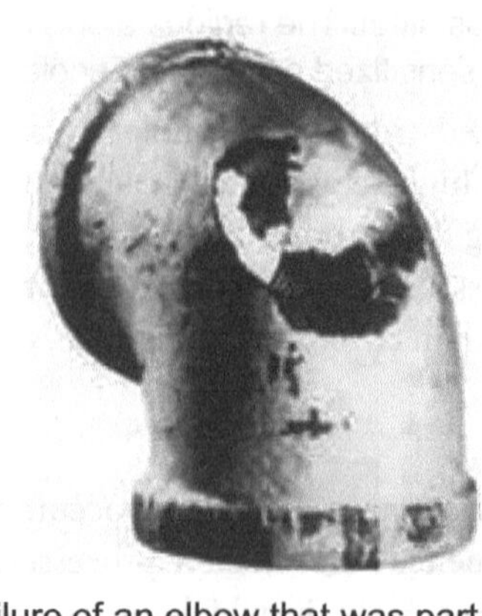

Fig. 5.12: Impingement failure of an elbow that was part of a steam condensate line.

One of the best ways to reduce erosion–corrosion is to change the design to eliminate fluid turbulence and impingement effects. Other materials may also be used that inherently resist erosion. Furthermore, removal of particulates and bubbles from the solution lessens its ability to erode.

5.4.8 Stress corrosion cracking

Stress corrosion cracking results from the combined action of an applied tensile stress and a corrosive environment; both influences are necessary. In fact, some materials that are inert in a particular corrosive medium become susceptible to this form of corrosion when a stress is applied. Small cracks form and then propagate in a direction perpendicular to the stress (Figure 5.13), which eventually fails. Failure behavior is characteristic of that for a brittle material, even though the metal alloy is intrinsically ductile. Furthermore, cracks may form at relatively low stress levels, significantly below the tensile strength. Most alloys are susceptible to stress corrosion in specific environments, especially at moderate stress levels.

The stress that produces stress corrosion cracking need not be externally applied; it could be a residual one that results from rapid temperature changes and uneven contraction or occur for two-phase alloys in which each phase has a different coefficient of expansion. Also, gaseous and solid corrosion products that are entrapped internally can give rise to internal stresses. The best measure to take to reduce stress corrosion is to lower the magnitude of the stress. This may be accomplished by reducing the external load or increasing the cross-sectional area perpendicular to the applied stress. Furthermore, an appropriate heat treatment may be used to anneal out any residual thermal stresses.

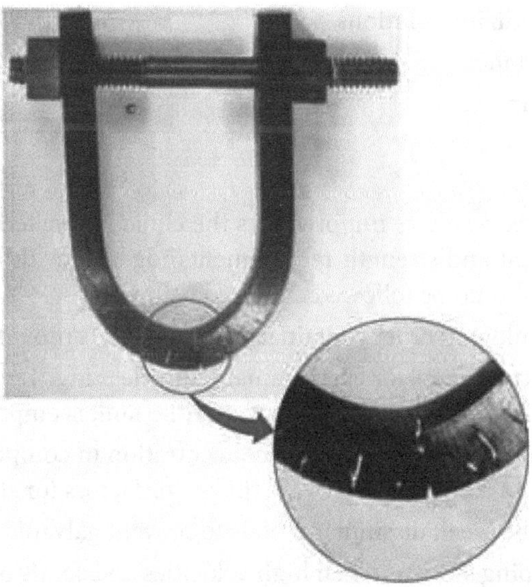

Fig. 5.13: A bar of steel bent into a horseshoe shape using a nut and-bolt assembly. While immersed in seawater, stress corrosion cracks formed along the bend at those regions where the tensile stresses are the greatest.

5.5 CORROSION CONTROL

The general methods of corrosion control include material selection, environmental alteration, design, coatings, and cathodic protection. These methods are briefly discussed below.

5.5.1 Materials selection

The most common method of preventing corrosion is the selection of the proper metal or alloy for a particular corrosive service. For instance, stainless steels have widespread application in resisting corrosion but it should be remembered that they do not resist all corrosives. In fact, under certain conditions, such as chloride containing mediums and stressed structures, stainless steels are less resistant than ordinary structural steel. Stainless alloys are more susceptible to localized corrosion such as intergranular corrosion, stress-corrosion cracking, and pitting attack than ordinary structural steels. Hence, stainless steels represent a class of highly corrosion resistant materials of relatively low cost that should be carefully used.

In alloy selection, there are several natural metal-corrosive combinations, which represent the maximum amount of corrosion resistance for the metal in that corrosive with least cost. Some of the natural combinations are as follows:

1. Stainless steels- nitric acid
2. Nickel and nickel alloys- caustic
3. Monel- hydrofluoric acid
4. Hastelloys- hot hydrochloric acid
5. Lead- dilute sulfuric acid
6. Aluminium- nonstaining atmospheric exposure
7. Tin- distilled water
8. Titanium- hot strong oxidizing solutions
9. Tantalum- ultimate resistance
10. Steel- concentrated sulfuric acid

5.5.2 Design

The design of a structure is frequently as important as the choice of materials of construction. Design should consider the mechanical and strength requirements together with an allowance for corrosion. Some of the design rules that should be followed are listed below:

(a) Weld rather than rivet tanks and other containers. Riveted joints provide sites for crevice corrosion.
(b) Design tanks and other containers for easy drainage and cleaning. Tank bottoms should be sloped toward drain holes so that liquids cannot collect after the tank is emptied.
(c) Avoid excessive mechanical stresses and stress concentration in components exposed to corrosive mediums. Mechanical and residual stresses are the major factors for stress-corrosion cracking.
(d) Avoid electrical contact between dissimilar metals to prevent galvanic corrosion.
(e) Avoid sharp bends in piping systems when high velocities and solids in suspension are involved.
(f) Avoid hot spots during heat-transfer operations. Heat exchangers and other heat-transfer devices should be designed to ensure uniform temperature gradients. Uneven temperature distribution leads to local heating and high corrosion rates.

5.5.3 Inhibitors

An inhibitor is a substance that, when added in small concentrations to an environment, decreases the corrosion rate of the metal. Inhibitors can be classified according to their mechanism and composition.

(a) **Adsorption Type Inhibitors**

These are the organic compounds which adsorb on the metal surface and suppress metal dissolution and reduction reactions. In most cases it appears that adsorption inhibitors affect both the anodic and cathodic processes. Typical example of this class of inhibitors are the organic amines.

(b) **Hydrogen Evolution Poisons**

Substances such as arsenic and antimony ions specifically retard the hydrogen-evolution reaction. As a result, these substances are very effective in acid solutions where hydrogen evolution occurs in the prominent cathodic reaction. However, these substances are ineffective in environments where other reduction processes such as oxygen reduction are the dominant cathodic reactions.

(c) **Scavengers**

These substances act by removing corrosive reagents from solution. Examples of this type of inhibitors are sodium sulfite and hydrazine, which remove dissolved oxygen from aqueous solutions as per the following reactions:

$$2Na_2SO_3 + O_2 \rightarrow 2Na_2SO_4$$

$$N_2H_4 + O_2 \rightarrow N_2 + 2H_2O$$

It is apparent that such inhibitors will work very effectively in solutions where oxygen reduction is the controlling corrosion cathodic reaction but will not be effective in strong acid solutions.

(d) **Vapor-Phase Inhibitors**

These substances are similar to the organic adsorption type inhibitors and possess a very high vapor pressure. As a consequence, these inhibitors can be used to inhibit atmospheric corrosion of metals without being placed in direct contact with the metal surface. During use, such inhibitors are placed in the vicinity of the metal to be protected and are transferred by sublimation and condensation to the metal surface. The vapor phase inhibitors are only effective if used in closed spaces such as inside packages or on the interior of machinery during shipment.

5.5.4 Coatings

Metallic coatings are used for corrosion prevention. If a metal coating is noble with respect to the substrate, it is necessary to avoid flaws in the coating such as cracks and pores. Such flaws could initiate corrosion, with the coating acting as the cathode and the substrate as the anode. As the exposed part of the anode at a crack or pore is very small, corrosion takes the form of a localized attack. Tin on steel is an example, exhibiting localized pinhole attack.

The tin coating on a steel substrate is produced either by dipping it in molten tin or by electroplating. The most common use of tinplate is for making food containers. The pinhole corrosion referred to above can take place on the outside of the container. Inside, many organic acids that are present in foodstuffs and fruit juices form complexes with tin. The concentration of the stannous ions is thereby lowered. The potential of tin decreases enough to make it anodic with respect to iron. In the absence of a suitable cathodic reaction, the corrosion rate is low.

If the metal coating is more active than the substrate (e.g., zinc or aluminium on steel), galvanic protection is offered to the substrate. The coating is anodic and corrodes first. Zn and Al, however, become passive after the initial attack. Galvanized iron (G.I.) is produced by dipping a low carbon steel sheet in molten zinc bath at about 450°C. On cooling in air, the zinc coating crystallizes forming 'zinc flowers'. Components such as buckets and drums made of galvanized iron are very suitable for aqueous environments. In the presence of oxygen, zinc hydroxide is precipitated as a protective layer. Aluminium coatings are deposited by the process of calorizing mainly to improve the oxidation resistance of steel. Decorative chromium plating is done over a first coating of nickel on automobile exterior components. Hard chromium plating produces a thicker, wear resistant surface.

Non-metallic coatings such as enamel, oil, paint and tar act by simply excluding water and oxygen and by providing a layer of high electrical resistance. For components used indoors, it is often enough to coat them with one layer of paint. For the protection of outdoor structures such as bridges, elevator cranes, harbour equipment, railway carriages, automobiles and bicycles, which are exposed to the action of the atmosphere, paints are of more complex composition. Also, their method of application is more intricate, involving several coatings.

5.5.5 Cathodic protection

Cathodic protection simply involves supplying electrons from an external source to the metal to be protected, making it a cathode. One cathodic protection technique employs a galvanic couple: the metal to be protected is electrically connected to another metal that is more active in the particular environment. The latter experiences oxidation and, upon giving up electrons, protects the first metal from corrosion. The oxidized metal is often called a sacrificial anode and magnesium and zinc are commonly used because they lie at the anodic end of the galvanic series. This form of galvanic protection for structures buried in the ground is illustrated in Figure 5.14(a).

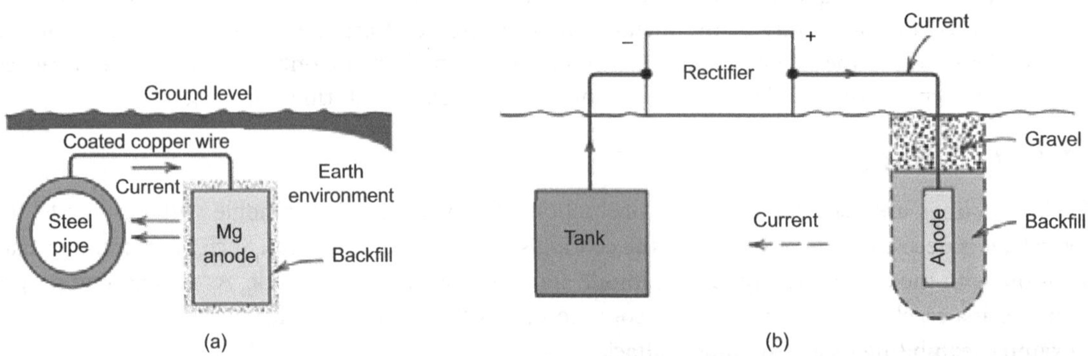

Fig. 5.14: Cathodic protection of (*a*) an underground pipeline using a magnesium sacrificial anode and (*b*) an underground tank using an impressed current.

For another method of cathodic protection, the source of electrons is an impressed current from an external dc power source, as represented in Figure 5.14(b) for an underground tank. The negative terminal of the power source is connected to the structure to be protected. The other terminal is joined to an inert anode (often graphite), which, in this case, is buried in the soil; high conductivity backfill material provides good electrical contact between the anode and the surrounding soil. A current path exists between the cathode and the anode through the intervening soil, completing the electrical circuit. Cathodic protection is especially useful in preventing corrosion of water heaters, underground tanks and pipes, and marine equipment.

5.6 SURFACE ENGINEERING AND HARDENING TREATMENTS

Surface engineering can be defined as the treatment of the surface and near-surface regions of a material to allow the surface to perform functions that are distinct from those functions demanded from the bulk of the material. It is intended to tailor the properties of the surfaces of engineering components so that their function and serviceability can be improved. The desired properties of surface engineered components include:

(a) Improved corrosion prevention through barrier or sacrificial protection.
(b) Improved oxidation and/or sulfidation resistance
(c) Improved wear resistance
(d) Reduced frictional energy losses
(e) Improved mechanical properties such as enhanced fatigue resistance or toughness
(f) Improved thermal insulation
(g) Improved aesthetic appearance

Engineering surfaces are often hardened, usually by thermal and thermochemical means, to improve their tribological and corrosion properties.

Thermal treatments involve the application of a specified thermal cycle to a material, in order to obtain desired mechanical and chemical properties. Common examples are the heating and quench hardening of steel, which are typically applied to change the iron–carbon distribution and modify the grain structure. Such treatments can be applied selectively by the local application of heat by induction, flame, electron beam or laser. Typically, surface layers hardened in this way can be up to several mm in depth.

Unlike thermal treatments, thermochemical treatments rely on modifying the existing surface rather than adding a coating on top of it. However, whereas a thermal treatment would utilize the existing carbon in steel, in a thermochemical treatment the composition of the surface is altered by adding one or more additional elements. This means that a much wider range of steels and other non-ferrous alloys could be treated.

The two main thermochemical treatment methods are carburizing and nitriding. The former is used, for example, to raise the carbon content in the near-surface regions of low carbon steels, and subsequent quenching can cause a martensitic transformation to occur, creating a hard surface with beneficial compressive stresses which can improve the fatigue strength. Typical case depths are up to about 4 mm. The main processes used are gas, vacuum and plasma carburizing. The last two have gained wide applicability due to the possibility of limiting the availability of oxygen at the surface. Carburizing is usually followed by a direct quench from the processing temperature, typically around 900°C. The addition of about 0.5% nitrogen in the process can allow lower process temperatures and reduces the case thickness, creating a carbonitrided surface.

Nitriding is a lower-temperature process typically carried out at around 500°C for steels. Gas nitriding and plasma nitriding are the most popular methods. The latter provides a shorter processing time and is claimed to give improved control over the nature of the surface layers. Case depths are typically 0.4–0.6 mm. Nitrocarburizing is a thermochemical process that can be applied to plain carbon, i.e. non-alloyed, steels. It is usually carried out at about 570 °C in ammonia with a carbon bearing gas and a carbonitride phase about 20 mm thick is formed on the surface in about 2 hours. Some porosity could occur, which is beneficial for oil retention, or even for impregnating with a sealing and/or lubricating medium. Table 5.1 gives some details of the main surface hardening treatments based on carbon and nitrogen.

5.6.1 Organic coatings

Painting and the application of various organic (plastic and rubber) coatings and linings are among the most widely used surface-engineering processes. Paints or linings that act as protective film to isolate the substrate from the environment exist in a number of different forms. Sheet linings, commonly of the vinyl or vinylidene chloride family are one such type of coating that can be either adhered to the surface to be protected or suspended as a bag within a tank. Hot applied organisols or plastisols can also be applied to a surface by dipping or flow coating to provide a protective film.

Powder coatings are being increasingly used to protect concrete-reinforcing rod, as pipeline coatings. Fine powders produced from high-molecular-weight resins of the thermoplastic vinyl and fluorinated hydrocarbon families or from thermoset resins of the epoxy and polyester families are applied to the surface to be protected by either electrostatic spray or fluidized-bed deposition. The metal being protected is usually preheated at the time of application and after application it is reheated to an elevated temperature (150-315°C).

Certain lining materials, such as hand lay-up fiberglass-reinforced plastics are also used to protect steel from corrosion. Such coating systems usually consist of an epoxy primer applied to a blast cleaned steel surface, followed by one or more polyester gel coats, along with layers of fiberglass veil as reinforcement. The system is then sealed with a layer of the polyester gel coat. Similarly, rubber lining is used to protect corrosion. There are various types of rubbers, but they can be categorized as prevulcanized or postvulcanized (vulcanized after application). Similarly, rubbers can be formulated with different hardness and chemical resistances. Commonly, a rubber lining is a composite of two or three different types of rubbers adhered to each other and to the surface.

Apart from proper application, the three main factors to consider for organic coatings are (a) surface preparation, (b) selection of primer or priming coat, and (c) selection of top coat. If the metal surface is not properly prepared, the paint may peel off because of poor bonding. If the primer does not have good adherence or is not compatible with the top coat, early failure occurs. Poor paint performance is in most cases due to poor application and surface preparation.

Surface preparation involves surface roughening to obtain mechanical bonding as well as removal of dirt, rust, mill scale, oil, grease, wax and other impurities. In other words, a clean, rough surface is needed. The best method is to grit-blast or sand blast the steel surface. Other methods are pickling, scraping, wire brushing, flame cleaning, chiselling and chipping. Other chemical methods are solvent degreasing, hot or cold alkali treatments, phosphatizing, chromate treatment, and electrochemical treatments such as anodizing and cathodic cleaning.

Table 5.1: List of thermomechanical treatments based on carbon and nitrogen.

Process	Description	Typical treatment temp. (°C)	Typical case depth (mm)	Typical surface hardness (HV)
Carburizing	A process in which a steel surface is enriched with carbon, at a temperature above the ferrite/austenite transformation. On subsequent quenching, an essentially martensitic case formed	850-950	0.25-4.0	700-900
Carbonitriding	Similar to carburizing, but involving nitrogen as well as carbon enrichment.	700-900	0.05-0.75	600-850
Nitrocarburizing	A process in which a steel or cast-iron surface is enriched with nitrogen, carbon and possibly sulphur at a temperature below the ferrite/austenite transformation	570	0.02-1.0	500-650
Nitriding	A process in which a steel surface is enriched with nitrogen, at a temperature below the ferrite/ austenite transformation.	500-525	0.4-0.6	800-1050

5.6.2 Electroplated coatings

Electroplating is defined as the deposition of a coating by electrolysis, that is, depositing a substance on an electrode immersed in an electrolyte by passing electric current through the electrolyte. The process can take place in an aqueous electrolyte near ambient temperatures or in a fused salt at high temperatures. Electrodeposits are applied to metal substrates for decoration, protection, corrosion resistance, chemical inertness, wear resistance, build up of substrate dimensions, electrical properties, magnetic properties, solderability, reflectance, and reduction of friction.

Examples of this application include the following:

(1) Tin or chromium plating (via continuous electrodeposition) of steel strip for food packaging and other container uses

(2) Electrogalvanizing or zinc plating of steel strip, sheet, stampings, forgings, wire, and screw machine products

(3) Zinc and cadmium plating of fasteners and other hardware items

(4) Bright nickel-chromium plating of household appliances, interior auto hardware, and standing ashtrays

(5) Chromium plating of gun bores

Aqueous solution electroplating provides decorative and protective finishes for use at ambient temperatures and in a variety of environments. The main drawback of electroplating is the inability of achieving uniform deposition, which is related to the throwing power of the electrolyte. *Throwing power* deals with the ability of the coating to serve uniformly in all areas of component even with irregular shape. Throwing power is dependent on the shape of the substrate, the anode/cathode configuration, current density, as well as on the composition and conductivity of the electrolyte. A second difficulty is that not all metallic elements can be deposited. Another problem with electroplating is evolution of hydrogen at the electrodes when cathode efficiency is less than 100%. If a ferrous substrate is to be plated, absorbed atomic hydrogen can cause embrittlement of the part. Unless the parts are heat treated to remove the absorbed hydrogen, they will be brittle and unusable for any application involving elastic strain.

Substrate preparation for plating is highly important to ensure good adhesion and surface quality. Maximum adhesion depends on both the elimination of surface contaminants in order to induce a metallurgical bond and the generation of a completely active surface to initiate plating on all areas. The cleaning steps for steel substrates usually involve precleaning, intermediate alkaline cleaning, electrocleaning and acid treatments.

Nickel plating: It is widely used for a corrosion and wear resistant finish. It also has a thin top coat of electrodeposited chromium. Typical applications include, decorative trim for automotive and consumer products and office furniture. Nickel deposits are also used for nondecorative purposes for improved wear resistance, for example on pistons, cylinder walls and balls studs.

Chromium electroplating: It is also used as decorative and hard coatings. Colored and tarnish-resistant chromium decorative coatings are produced over a base deposit of copper and nickel for applications such as those noted above for nickel. Hard chromium coatings are used for hydraulic pistons and cylinders, piston rings, aircraft engine parts, and plastic molds, where resistance to wear, heat abrasion, and corrosion are required.

Cadmium and zinc electroplating: It provides galvanic corrosion protection when coated on steel. Deposit thickness can vary between 5 and 25 μm. Cadmium is preferred for the protection of steel in marine environments, whereas zinc is preferred in industrial environments. Cadmium is also preferred for fastening hardware and connectors because its coefficient of friction is less than that of zinc. Cadmium is toxic and should not be used in parts that would have contact with food. Precautions for minimizing hydrogen embrittlement should be taken because cadmium plating is more susceptible to such embrittlement than any other plated metal.

5.6.3 Electroless nickel plating

Electroless nickel coatings are produced by the autocatalytic chemical reduction of nickel ions from an aqueous solution. They do not require electrical current to be produced. Two types of electroless coatings are produced: nickel-phosphorus and nickel-boron alloy coatings. Both coating types are used in applications requiring a combination of corrosion and wear resistance.

Nickel-phosphorus alloy coatings: these coatings are classified as either low phosphorus (1 to 4 wt% P), medium phosphorus (5 to 8% P), or high phosphorus (9 to 12% P). These amorphous barrier coatings provide excellent corrosion protection in a wide range of environments. The resistance to attack in neutral and acidic environments is increased as the phosphorus content is increased in the deposit.

The reverse is true in alkaline corrosive environments.

Nickel-boron alloy coatings: these coatings typically contain 5% boron and have excellent resistance to wear and abrasion. They, however, are not completely amorphous and have reduced resistance to corrosion; furthermore, they are much more expensive than nickel phosphorus coatings.

5.6.4 Thermal spray coatings

Thermal spraying comprises a group of processes in which finely divided semi-molten metallic or non-metallic material is sprayed onto a prepared substrate to form a coating. The sprayed material is originally in the form of wire or powder. As the coating materials are fed through the spray unit, they are heated to a partly molten (mostly surfaces) state and propelled by a stream of compressed gas onto the substrate. As the particles strike the surface, they flatten and form thin platelets that conform and adhere to the irregularities of the prepared surface and to each other. They cool and accumulate, particle by particle, into a lamellar, castlike structure. The spray gun generates the necessary heat for melting through combustion of gases, electric arc, or a plasma.

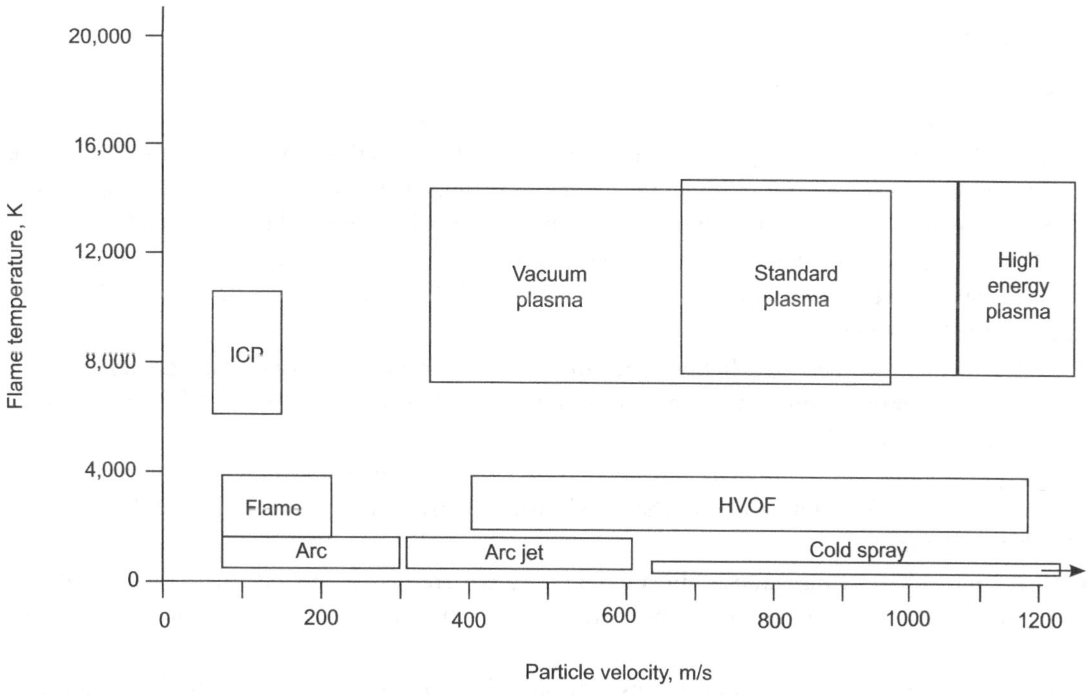

Fig. 5.15: The temperature-velocity distribution envelope of thermal spray processes. HVOF (high velocity oxyfuel); ICP (induction coupled plasma).

Thermal spray (TS) is a generic term used to describe a group of processes, including flame spraying, plasma spraying, arc metallization, detonation gun (D gun), high velocity oxyfuel (HVOF), and cold spray, that can be used to apply a variety of different coating materials for corrosion protection. Figure 5.15 depicts the temperature-velocity processing envelope that allows materials engineers to create a vast array of coatings for applications that include coatings for orthopaedic hips, thermal barrier coatings for aerospace components, and coatings to protect against corrosion.

Table 5.2 lists different thermal spray processes. Those that are based on similar technological principles or that have equivalent names are grouped onto the same lines. The choice of process or material depends on the engineering application.

Table 5.2: Thermal spray process parameters

Process	Temperature (°C)	Velocity (m/s)
Oxyacetylene gas flame spraying	3000	150
Atmospheric plasma spraying	12000	900
Low-pressure plasma spraying	1200	900
Detonation gun spraying	300	1000
Hypersonic plasma spraying, high velocity oxyfuel	3000	800
Arc metallization, wire spraying	1500	200

5.6.5 Conversion Coatings

Phosphate coating is the treatment of iron, steel, galvanized steel, or aluminum with a dilute solution of phosphoric acid and other chemicals in which the surface of the metal, reacting chemically with the phosphoric acid media, is converted to an integral, mildly protective layer of insoluble crystalline phosphate. Phosphate conversion coatings are applied to various metal substrates to enhance corrosion resistance, increase paint or plating adhesion, or both. Other applications include the use of phosphate conversion coatings for their electrical insulating properties (for example, in electric motor laminations) and for lubricity (for example, to increase the formability of sheet metals). The basic types of phosphate coatings are the following:

(a) **Iron phosphates**—these are lightweight, amorphous phosphate coatings that do not contain significant amounts of divalent metal ions from solution. Coating weights range from 0.16 to 0.80 g/m^2 (15 to 75 mg/ft^2).

(b) **Zinc phosphates**—these are medium-weight, crystalline phosphate coatings that contain divalent metal ions from the solution and/or the metal surface. Coating weights range from 1.4 to 4.0 g/m^2 (130 to 370 mg/ft^2).

(c) **Heavy phosphates (manganese phosphates)** — these are heavy coatings that contain divalent metal ions from solution and from the metal surface. Coating weights range from 7.5 to 30 g/m^2 (700 to 2800 mg/ft^2).

There are two major categories for the uses of phosphate coatings: bare corrosion protection and painted corrosion protection. Bare phosphate simply means that the phosphate coating is not painted. Because of the crystalline nature of zinc phosphates and manganese phosphates, these coatings can hold oils and waxes so well that the corrosion resistance of the metal is increased much more than is expected based on either the phosphate or the oil separately. Similarly, the phosphate coating holds paint physically. The basic process in the formation of any phosphate coating is the precipitation of a divalent metal and phosphate ions (PO_4^{3-}) on a metal surface. Phosphate salts, particularly divalent metal salts, are soluble in acid solutions and insoluble in neutral and basic solutions. The phosphate baths are acidic enough to keep the ions in solution. When metal is exposed to the solution, the acid attacks the metal surface. Two changes occur in the solution directly adjacent to the metal surface. First the acid is neutralized and the pH rises. Second, the concentration of metal ions increases. The two methods of applying phosphate conversion coatings are spraying and immersion. In general, spray coating is less expensive, but immersion coating provides better quality. Immersion phosphating produces changes in the phosphate coating composition that result in better painted performance with some paint systems.

Chromate conversion coatings are formed by a chemical or an electrochemical treatment of metals or metallic coatings in solutions containing hexavalent chromium (Cr^{6+}) and, some other components. The process results in the formation of an amorphous protective coating composed of the substrate, complex chromium compounds, and other components of the processing bath. Chromate conversion coatings are applied primarily to enhance bare or painted corrosion resistance, to improve the adhesion of paint or other organic finishes, and to provide the metallic surface with a decorative finish. Chromating processes are widely used to finish aluminum, zinc, steel, magnesium, cadmium, copper, tin, nickel, silver, and other substrates. They are most frequently applied by immersion or spray coating, but other methods of application, such as brushing, roll coatings, dip and squeegee, electrostatic spraying, or anodic deposition, are used in special cases. The disposal of spent solutions and rinse waters requires waste treatment. Hexavalent chromium must be reduced to Cr^{3+} before neutralizing and precipitation. Sodium pyrosulfite ($Na_2S_2O_5$) is usually used as the reducing agent in smaller operations, while for larger plants, sulfur dioxide (SO_2) is preferred for economic reasons.

Wastewater treatment sludges from chromating operations are considered hazardous waste. As a result, the use and disposal of chromium and chromium compounds have received much regulatory attention because of the toxicity of chromium and indications that it is a cancer-causing agent. Possible elimination of chromate conversion coatings due to regulatory restrictions is particularly troublesome for the aircraft industry. In applying the coating to the entire aircraft aluminium structure, the subsequent rinse process can generate large quantities of chromium containing wastes. The challenges of adequately maintaining aging aircraft will help drive the search for effective chromate substitutes.

5.6.6 Deposition Process

The vapor-deposition processes fall into two major categories—physical vapor deposition (PVD) and chemical vapor deposition (CVD). All these processes have one characteristic in common: the deposition species are transferred and deposited onto the substrate in the form of individual atoms or molecules. In this respect, they are fundamentally different from particulate- or liquid-deposition processes, such as thermal spraying or electroplating. A primary benefit of vapor deposition is that it does not cause threat to the environment, because there are no solvents or chemicals to dispose of. With the current awareness of environmental, safety, and health considerations, this crucial factor favors vapor deposition over other processes that could be more harmful to the environment and health.

Physical vapor deposition: Sputtering is the principal PVD process. It involves the transport of a material from a source (target) to a substrate by bombardment of the target by gas ions that have been accelerated by a high voltage. Atoms from the target are ejected by momentum transfer between the incident ions and the target. These ejected particles move across the vacuum chamber to be deposited on the substrate. Figure 5.16 is a schematic of the sputtering process. The process occurs in the presence of an inert gas at low pressure (10^{-3} to 0.1 torr); in most cases, the gas is argon. Argon has higher mass than other noble gases, such as neon or helium, and is easier to ionize.

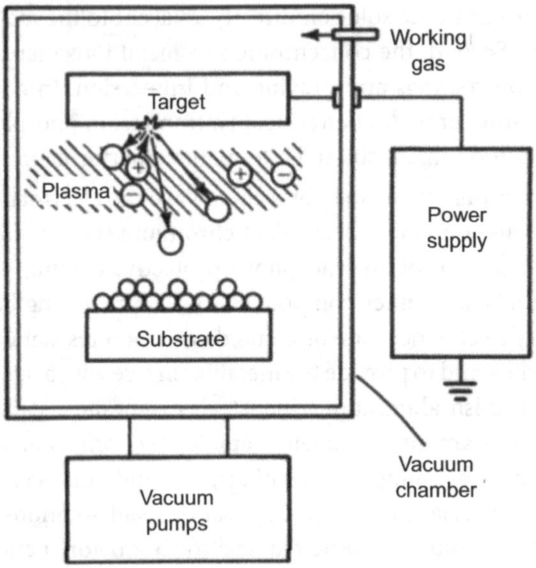

Fig. 5.16: Basic sputtering process

Higher mass gives a higher sputtering yield, especially if the mass of the bombarding particle is the same order of magnitude or greater than the mass of the target atom. Other gases, such as oxygen or nitrogen, can also be used, but they could react chemically with the target. The sputtering process begins when an electric discharge is produced and the argon becomes ionized. The low-pressure electric discharge is known as glow discharge, and the ionized gas is termed plasma. The argon ions hit the solid target, which is the source of the coating material.

The target is negatively biased and therefore attracts the positively charged argon ions, which are accelerated in the glow discharge. This attraction of the ions to the target (also known as bombardment) causes the target to sputter, which means that material is dislodged from the target surface because of momentum energy exchange. The higher the energy of the bombarding ions, the higher the rate of material dislodgement.

For the deposition of thin metal films, sputtering is the best technique. Deposition is feasible in a controllable manner for both compound and elemental targets (the source of the coating material). Adhesion is good and can be further improved by sputter cleaning the substrate or by bias sputtering. Large equipment is readily available, and sputtering has become a highly automated process. The quality, structure, and uniformity of the deposit are excellent. The disadvantages of sputtering include its thickness limitation, the need for a vacuum, and high cost. It is also a line-of-sight process that requires special fixtures and shaped targets to coat uniformly. The substrate being processed must be capable of withstanding processing temperatures of 260 to 540 °C.

Chemical vapor deposition is a well-established deposition process that is used extensively in the semiconductor and cutting-tool industries. Figure 5.17 is a schematic of thermal CVD equipment. It has the unique ability to deposit thick, dense, high-quality films (up to 6.4 mm thick or more) at a cost that is generally lower than that of PVD processes. The high temperature, about 1000 °C, necessary for CVD promotes good adhesion, but at the same time restricts the type of substrate that can be coated to ceramics, refractory metals, and special alloys. In most cases, steel will require heat treatment in a vacuum or protected atmosphere after CVD coating and will often change dimensions sufficiently to require postdeposition machining if close tolerances are required. Another constraint resulting from the high temperature of deposition is the stress in the deposit due to the difference in thermal expansion between substrate and deposit.

These stresses may be sufficient to cause cracking, spalling, and loss of adhesion. The technique of plasma CVD can be used at much lower temperatures, for example, semiconductor applications. The materials that can be easily deposited by CVD are more limited than with PVD. Chemical vapor deposition is particularly useful for depositing compounds and refractory metals and exhibits excellent corrosion resistance.

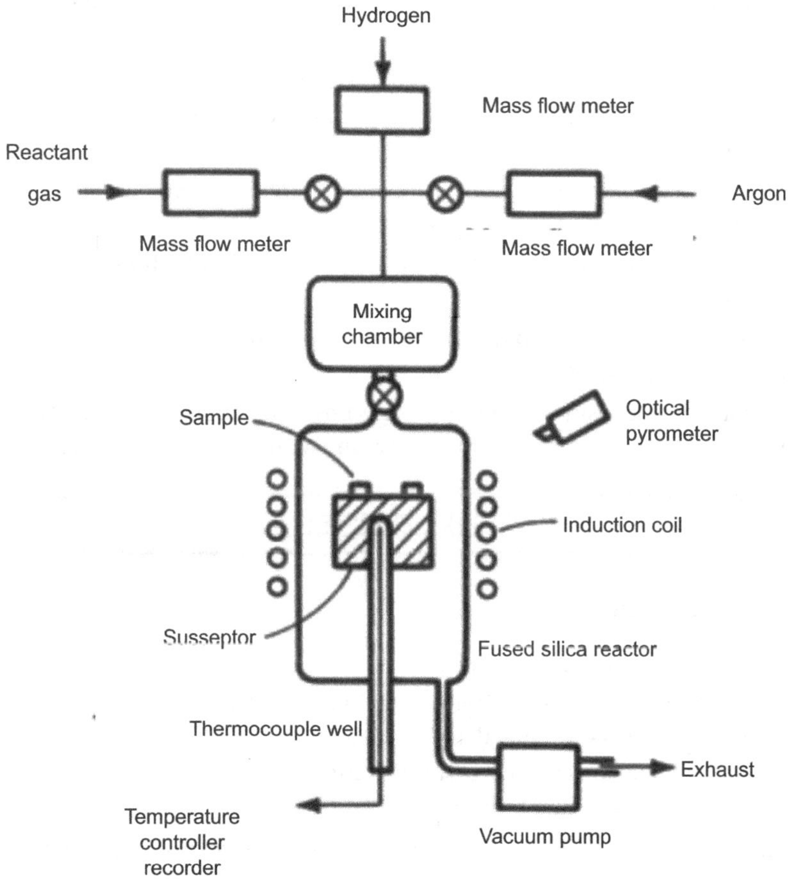

Fig. 5.17: Thermal CVD reactor

The reactive gases are introduced in the mixing chamber. They then pass into the deposition chamber and contact the heated part to be coated (sample); at this point, they react and deposition occurs. By-products are removed through the vacuum system. Chemical vapor deposition reactions can occur over a range of pressures. Low pressure increases boundary layer diffusion and uniformity, often at the expense of deposition efficiency. Each application and each reaction must be analysed to determine the optimal conditions for deposition.

UNIT SUMMARY

1. The present chapter discusses the significance of corrosion, its importance with relevance to engineering applications and its impact on the economy.
2. Anode, cathode, electrolyte, and metallic path are the different components of a corrosion cell and various factors such as oxidizing/reducing environment, velocity, temperature and metallurgical factors are responsible to initiate corrosion.
3. Metallic corrosion is typically electrochemical, involving both oxidation and reduction reactions.
4. The eight different forms of corrosion are explained in detail with illustrative examples and mechanisms.
5. The different corrosion control methods could be broadly classified based on proper selection of materials, design, altering the environment using inhibitors, protective coatings and application of cathodic protection.
6. The important aspect of surface engineering and different surface hardening treatments such as carburizing and nitriding are briefly explained.
7. The different organic and metallic coatings, surface preparation for these coatings, application and limitation have also been discussed.
8. The concepts of electroplating and electroless plating, thermal spray coating and methods of deposition are also explained.
9. Finally, the different deposition methods using physical vapor deposition (PVD) and chemical vapor deposition (CVD) have been explained along with the working principles in this chapter.

EXERCISES

Multiple Choice Questions

1. What is the process of metallic corrosion?
   (a) Formation of a protective oxide layer on the metal surface
   (b) Chemical reaction between metal and oxygen
   (c) Dissolution of metal ions in a corrosive medium
   (d) Mechanical wear of the metal surface

2. Which of the following factors does not influence the rate of corrosion of a metal?
   (a) Temperature (b) Humidity
   (c) Exposed surface area (d) Density of the metal

3. Rust is an example of corrosion that specifically affects:
   (a) Aluminum (b) Copper
   (c) Iron (d) Zinc

4. Which type of corrosion occurs when two dissimilar metals are in electrical contact in the presence of an electrolyte?
   (a) Pitting corrosion (b) Galvanic corrosion
   (c) Crevice corrosion (d) Uniform corrosion

5. Which of the following case is considered as technological advantage of corrosion
   (a) Synthesis of printed circuit boards (b) Anodizing
   (c) Development of microstructure in steel (d) All of these
6. Which corrosion control method does not involve the use of chemicals that inhibit the corrosion process?
   (a) Cathodic protection (b) Anodic protection
   (c) Corrosion inhibitors (d) Surface passivation
7. Which corrosion control method involves applying a sacrificial metal to protect the base metal from corrosion?
   (a) Cathodic protection (b) Anodic protection
   (c) Passivation (d) Galvanization
8. Corrosion of metals can be accelerated by which of the following?
   (a) Decreasing the surface area (b) Increasing the temperature
   (c) Increasing the pH level (d) Increasing the presence of corrosive ions
9. Rust is an example of corrosion that specifically affects:
   (a) Aluminum (b) Copper
   (c) Iron (d) Zinc
10. Which of the following is a common material used for anodizing?
    (a) Zinc (b) Copper
    (c) Aluminum (d) Steel
11. Which of the following is suitable methods for corrosion protection of mild steel
    (a) Phosphate coating (b) CVD
    (c) Anodizing (d) PVD
12. Which of the following is a common material used for electroplating?
    (a) Silver (b) Glass
    (c) Nickel (d) Ceramic
13. Which of the following is a common purpose of applying coatings to a surface?
    (a) Corrosion protection (b) Improved service life
    (c) Enhanced electrical conductivity (d) Increased wear resistance
14. Which of the following is not a common method for applying coatings?
    (a) Spraying (b) Dipping
    (c) Brushing (d) Welding

ANSWERS															
1.	(c)	2.	(d)	3.	(c)	4.	(c)	5.	(d)	6.	(a)	7.	(a)	8.	(b)
9.	(c)	10.	(c)	11.	(a)	12.	(a)	13.	(a)	14.	(d)				

Short and Long Answer Type Questions

1. Define corrosion and write down five common examples.
2. State various forms of corrosion with at least one example of each.

3. What is the difference between pitting corrosion and crevice corrosion?
4. How does corrosion differ from erosion?
5. Explain the mechanism of electrochemical corrosion.
6. Why does corrosion generally occur at the anode?
7. What determines whether a given area is anodic or cathodic?
8. What is the limitation on the use of the electromotive series in the corrosion of metals?
9. Why do aluminium and chromium show superior corrosion resistance than iron?
10. What is meant by anodizing? How does it affect corrosion resistance?
11. What three factors are necessary to form a galvanic cell?
12. Explain and describe the mechanism of dezincification.
13. Discuss the differences and similarities between intergranular corrosion and stress corrosion?
14. Describe each of the following: galvanic corrosion, pitting corrosion, crevice corrosion, cavitation corrosion, stress corrosion.
15. Explain cathodic and anodic protection of metal/alloys.
16. List seven methods of corrosion protection. Explain why and when each method is effective.
17. Explain the following processes with an example of their application: electroplating, anodizing, chemical vapour deposition, physical vapour deposition.
18. Why stainless steel is stainless and has very high corrosion resistance?
19. Write down the factors affecting corrosion of metals/alloys?
20. Why ceramic rotor shows superior corrosion resistance than a metallic rotor?
21. A foundry person seeks your suggestion to improve the corrosion resistance of plain carbon steel. What type of alloying elements will be suggested by you and why?

PRACTICAL

Experiment title: To study the corrosion behaviour of mild steel, stainless steel and aluminium in 3.5%NaCl solution.

Objectives:

(a) To perform the immersion test of mild steel, stainless steel and aluminium in 3.5%NaCl solution for a period of 30 days.

(b) Determine the corrosion rate of the different samples after 30 days of the immersion test.

Apparatus/Instrument: Anhydrous sodium chloride, conical flask, distilled water, weighing balance, mild steel, stainless steel and aluminium samples.

Theory:

Students can refer to ASTM G31-21, which describes the standard guide for performing the immersion corrosion testing of metals and determination of corrosion rate.

Procedure:

Students can refer to ASTM E3-11 to understand the method for metallographic sample preparation for immersion test.

Result: The parameters obtained after the immersion test needs to be tabulated

Sample	Initial weight (g)	Final weight (g)	Area (cm2)	Density (g/cm3)	Corrosion rate (mm/y)
Mild steel					
Stainless steel					
Aluminium					

Conclusion:

The corrosion rate of mild steel, stainless steel and aluminium after 30 days of the experiment is ....

KNOW MORE

Students can refer to advanced concepts on corrosion engineering such as Modern mixed potential theory, electrochemical impedance spectroscopy and methods for corrosion monitoring and control.

REFERENCES AND SUGGESTED READINGS

1. Fontana, G. Mars, Greene and D. Norbert, Corrosion Engineering, McGraw-Hill, New York, 1967.
2. A.J. Bard, L.R. Faulkner, Electrochemical Methods: Fundamentals & Applications, John Wiley & Sons Inc. New York, 1980.
3. Z. Ahmad, Principles of Corrosion Engineering and Corrosion Control, Elsevier Science & Technology Books, 2006.
4. H.H. Uhlig, R.W. Revie, Corrosion and Corrosion Control (3 rd Ed), Jonh Wiley & Sons Inc. New York, 1985.

Dynamic QR code for further reading

CO AND PO ATTAINMENT TABLE

Course outcomes (COs) for this course can be mapped with the programme outcomes (POs) after the completion of the course and a correlation can be made for the attainment of POs to analyze the gap. After proper analysis of the gap in the attainment of POs necessary measures can be taken to overcome the gaps.

Table for CO and PO attainment

Course Outcomes	ATTAINMENT OF PROGRAMME OUTCOMES (1- Weak Correlation; 2- Medium correlation; 3- Strong Correlation)						
	PO-1	PO-2	PO-3	PO-4	PO-5	PO-6	PO-7
CO-1	3	2	2	1	3		
CO-2	3	2	2	1	3		
CO-3	3	2	2	1	3		
CO-4	3	2	2	1	3		
CO-5							

The data filled in the above table can be used for gap analysis.

INDEX

I

L

M

N

O

P

□□